EFFECTS OF WETLAND CONVERSION TO FARMING ON WATER QUALITY AND SEDIMENT AND NUTRIENT RETENTION IN A TROPICAL CATCHMENT

ABIAS UWIMANA

Thesis committee

Promotor
Prof. Dr K.A. Irvine
Professor of Aquatic Ecosystems
IHE Delft Institute for Water Education
Wageningen University & Research, Aquatic Ecology and Water Quality Management

Co-promotor
Dr A.A. van Dam
Associate Professor of Environmental Systems Analysis
IHE Delft Institute for Water Education

Other members
Prof. Dr J. Wallinga, Wageningen University & Research
Prof. Dr P. Haygarth, University of Lancaster, UK
Prof. Dr M.J. Wassen, Utrecht University
Prof. Dr L. Triest, Vrije Universiteit Brussel, Belgium

This research was conducted under the auspices of the SENSE Research School for Socio-Economic and Natural Sciences of the Environment

EFFECTS OF WETLAND CONVERSION TO FARMING ON WATER QUALITY AND SEDIMENT AND NUTRIENT RETENTION IN A TROPICAL CATCHMENT

Thesis
submitted in fulfilment of the requirements of
the Academic Board of Wageningen University and
the Academic Board of the IHE Delft Institute for Water Education
for the degree of doctor
to be defended in public
on Thursday, 28 November 2019 at 01:30 p.m.
in Delft, the Netherlands

by Abias Uwimana
Born at Kora, Rwanda

CRC Press/Balkema is an imprint of the Taylor & Francis Group, an informa business

Published by:
CRC Press/Balkema
Schipholweg 107C, 2316 XC, Leiden, the Netherlands
Pub.NL@taylorandfrancis.com
www.crcpress.com – www.taylorandfrancis.com
ISBN 978-0-367-85973-2 (Taylor & Francis Group)
ISBN 978-94-6395-073-2 (Wageningen University)
DOI https://doi.org/10.18174/497779

In memory of my Father Hakizimana Jonas and Grand Brother Niyonzima Fanuel who passed away very early.

Hoping to meet again in the morning for resurrection.

Acknowledgements

This PhD research resulted from the implementation of the Water Resources and Environmental Management (WREM) capacity building project (NPT/RWA/051) between the University of Rwanda (former National University of Rwanda) and IHE-Delft Institute for Water Education (former UNESCO-IHE, Delft, The Netherlands). The project, which was implemented between 2005 and 2011, aimed to contribute to poverty alleviation and sustainable socio-economic development in Rwanda by stimulating solution-oriented research related to water resources and environmental management.

I would like to express my gratitude to NUFFIC and the Dutch government for the financial support for this research, and to IHE-Delft and University of Rwanda (UR) for providing the technical research facilities. I would like to extend my sincere thanks to the whole supervising team for the scientific guidance throughout this research. My research started in 2009 under the supervision of Prof. Jay O'Keeffe as promoter and Dr. Anne van Dam as co-promoter. One year later, Prof. Kenneth Irvine took over as promotor and Dr. Gretchen Gettel also joined the supervisory team. I am highly indebted to Dr. Anne van Dam for the day-to-day guidance and help with statistical analyses. I am also grateful to Prof. Dr. Eng. Umaru Garba Wali, the Dean of School of Engineering, University of Rwanda for his tireless advice and encouragement throughout this study. Your kindness and generous behaviour will always be remembered. I would also like to acknowledge the contribution by UR MSc-students Bigirimana Bonfils and Nyirimigabo Alain in field data collection, Mr. Nkundimana Emmanuel for chemical analysis and Uwayezu Ernest for GIS analysis. Your contribution made this work successful.

I am indebted to my field research assistant Mugisha Jean de Dieu for his professionalism in the work. We energetically worked together in burning sunlight, heavy rainfall and even on occasion spent the night together in the field.

Special thanks to Dr. Erik de Ruyter (Project Director, WREM Project on the side of IHE Delft), Prof. Dr. Innocent Nhapi (Project Coordinator, WREM), and Dr. Karangwa Aphrodis (former Manager of the WREM Project), for their supportive roles during my PhD application and enrolment. Special thanks to Ms. Jolanda Boots (PhD Fellowship Officer at IHE Delft) for the nice cooperation in the course of this research. Many thanks to the Society of Wetlands Scientists (SWS) for the financial support to attend the 2016 SWS Annual Meeting in Corpus Christi, Texas (USA).

Last but not least, I would like to thank the members of my family, my mother Nyirabagwiza Adele, my wife Nimurere Solange, my children Irasubiza Geila, Irumva Elsa and Igiraneza Jessy, my sister Mukabera Thabia and my brothers Bizimana William and Nzarora Dan for their deep love, care and prayers, which provided me with the necessary moral support, comfort and energy to successfully complete this challenging venture. You missed me for a long time and this thesis is the fruit of your patience and encouragement.

Abias Uwimana
IHE Delft Institute for Water Education, Delft, The Netherlands
June 2019

Summary

Agricultural intensification may accelerate rapid loss of wetlands, increasing the concentration of nutrients and sediments in water bodies. Diversification and co-existence of wetland provisioning and regulating services can help to provide food security while at the same time enhancing sediment and nutrient retention. The study objective was to assess the effects of conversion of natural wetlands to agricultural farms on the retention of sediments, nitrogen and phosphorus. Specific objectives were to: (1) assess the effects of discharge and land use and land cover (LULC) on the water quality; (2) assess the effects of agricultural land use on sediment and nutrient retention; (3) assess the effects of conversion of wetlands to fish and rice farming on water quality; and (4) assess the effects on conversion of wetlands to fish and rice farming on sediment and nutrient pathways. To achieve the study objectives a combination of landscape-scale synoptic surveys (catchment, reaches) and mesocosm surveys (experimental plots) was used to study relationships between land use and land cover (agricultural farms, fishponds/dam, forest/wetlands), landscape features (slope, area, length, population density), rainfall, water discharge and water quality parameters (nitrogen, phosphorus, suspended solids, dissolved oxygen, conductivity, pH and temperature).

Results from the catchment study showed seasonal trends in water quality associated with high water flows and farming activities. Across all sites, total suspended solids (TSS) related positively to discharge, increasing 2-8 times during high flow periods. Electrical conductivity (EC,) temperature, dissolved oxygen (DO) and pH decreased with increasing discharge, while total nitrogen (TN) and total phosphorus (TP) did not show a clear pattern. TSS concentrations were consistently higher downstream of reaches dominated by rice and vegetable farming than in reaches dominated by forest/wetland and fishponds/dam. For TN and TP, results were mixed, but suggested concentrations of TN and TP increased during the dry and early rainy (and farming) season, and then washed out during the rainy season, with subsequent lower concentration at the end of the rains owing to dilution. Rice and vegetable farming generate greater amounts of transported sediment compared with ponds/reservoir and grass/forest. Valley bottoms dominated by grass/forest and ponds/reservoir types were generally associated with positive net yields of nutrients and sediments, while those with agricultural land covers had a net negative yield, resulting in net export. Water was retained only in reaches with ponds/reservoirs owing to their characteristics and management requirement to store water for the subsequent farming activities Seasonally, there was a strong relationship between net yield and discharge, with 93%, 60% and 67% of the annual TSS, TP and TN yields, respectively, transported during 115 days with rain. During low flow periods, all LULC types had positive net yields of TSS, TP and TN (suggesting retention), but during high flow periods had negative net yields (suggesting export). Significant effects of hillside land use on sediment and nutrient yields were also found.

Results from the mesocosm studies in the experimental plots showed that fish farming used about 2 times less water than rice farming (average of 101 mm m^{-2} d^{-1} compared with 191 mm m^{-2} d^{-1}). Biomass increase in fish farming was much lower (30 g m^{-2} in 8-months) than in rice (2500 - 4500 g m^{-2} in 4 months) and wetland plots (1300 - 1600 g m^{-2} in 8-11 months). Over the two seasonal periods studied, higher concentrations of TSS, TP and TN in inflows and outflows were associated mainly with human activities (cleaning of water supply canals, rice plot ploughing, weeding and fertilizer application, and fishpond drainage and dredging). As a result, fishponds and rice plots generally had consistently higher TSS concentrations in surface outflows (5-9506 and 4-2088 mg/L for fishponds and rice plots, respectively) than inflows (7-120 and 5-2040 mg/L for fishponds and rice plots, respectively). For TN and TP, results were more mixed, but with peaks of concentrations associated with the periods of land ploughing, weeding and fertilizer application, and fishpond drainage, and dredging. In wetland plots, TSS and TN significantly decreased from the inlet to the outlet, owing to the absence of disturbances of the plots and probably other mechanisms (higher settling/adsorption, nutrient uptake and denitrification).

Ploughing and weeding during the first three months of rice farming, and water renewal and dredging in the middle and end period of fish farming generate and discharge large amounts of sediment, N and P in the outflow of these systems. This makes fishponds a temporal sediment and nutrient storage during early farming stages and a source towards the end of farming. In contrast, rice farming generates sediments and nutrients early during the farming period (ploughing, weeding, transplantation and fertilizer application) and traps them towards the end. Despite higher fertilizer input in rice farms, N and P storage in soil decreased in rice farms (by 4.7 and 1.4%, respectively), but increased in fishponds (by 3.3% and 4.4%) and wetlands (3.8% and 1%). The decrease in nutrient soil storage was attributed to higher N and P uptake in rice plots (on average 662 and 270 mg m^{-2}d^{-1} of N and P, respectively) than in wetlands (359 and 121 mg m^{-2} d^{-1} of N and P) and fishponds (7.4 and 4.4 mg m^{-2} d^{-1} of N and P). To improve people's livelihoods and economic development while maintaining the water quality downstream the farming area, it is very important to consider the efficient recycling of water, sediments and nutrients through rotational rice crops in fishpond sediments. Wetlands could be more widely integrated with rice and fish farming to act as sediment and nutrient buffers in the critical periods of faming (rice farm ploughing and weeding, and pond drainage). This could lead to more sustainable farming with less erosion and reduced sediment and nutrient loads downstream.

Samenvatting

Intensivering van de landbouw kan het verlies en de achteruitgang van wetlands bespoedigen en leiden tot verhoogde concentraties van nutriënten en sediment in oppervlaktewater. Diversificatie en een goede balans tussen de producerende en regulerende ecosysteemdiensten kunnen bijdragen aan voedselzekerheid en tegelijkertijd de retentie van sediment en nutriënten bevorderen. Het doel van deze studie was om de effecten te bestuderen van het omzetten van natuurlijke wetlands in landbouwgrond op de retentie van sediment, stikstof en fosfor. De studie werd uitgevoerd in het stroomgebied van de Migina rivier in zuidelijk Rwanda, waar veel wetlands in valleibodems worden gebruikt voor landbouw. Sub-doelen waren: (1) beoordelen van het effect van rivierafvoer en landgebruik op de waterkwaliteit; (2) beoordelen van de effecten van landgebruik op de retentie van sediment en nutriënten; (3) beoordelen van de conversie van wetlands naar rijstteelt en visvijvers op de waterkwaliteit; en (4) beoordelen van het effect van rijstteelt en visvijvers op de sediment- en nutriëntenstromen. Hiervoor werd een combinatie van studies op landschapsniveau (stroomgebied, riviertraject) en met mesocosms (experimentele systemen) gebruikt om de relaties te onderzoeken tussen landgebruik (landbouw, visvijvers, reservoirs, bos, wetlandvegetatie), landschapskarakteristieken (helling, oppervlakte, lengte, bevolkings-dichtheid), regenval, rivierafvoer, en waterkwaliteitsparameters (stikstof, fosfor, zwevende stoffen, zuurstof, geleidbaarheid, zuurgraad en temperatuur).

De studie op landschapsniveau resulteerde in patronen in waterkwaliteit die verband hielden met hoge waterafvoer en landbouwactviteit. Op alle onderzochte locaties was er een positieve relatie tussen de concentratie zwevende deeltjes (TSS) en de rivierafvoer, met 2-8 keer hogere concentraties tijdens perioden met hoge afvoer. De geleidbaarheid (EC), temperatuur, zuurstofconcentratie (DO) en zuurgraad (pH) van het water daalden met verhoogde afvoer, terwijl de concentraties stikstof (TN) en fosfor (TP) geen duidelijk patroon vertoonden. TSS concentraties waren altijd hoger benedenstrooms van rivier-trajecten met rijst- en groententeelt, dan bij trajecten met bos- of wetlandvegetatie of bij visvijvers of reservoirs. Voor TN en TP waren de resultaten gemengd, maar in het algemeen waren de concentraties verhoogd tijdens het droge seizoen en aan het begin van het regen- en landbouwseizoen, en lager tijdens het regenseizoen vanwege verdunning en uitspoeling. Rijst- en groententeelt genereerden grotere hoeveelheden sediment dan visvijvers, reservoirs of bos- en wetlandvegetatie. Valleibodems met overwegend gras- en bosvegetatie of vijvers of reservoirs hadden over het algemeen positieve netto sediment- en nutriëntenopbrengsten (retentie), terwijl valleibodems met landbouw een negatieve opbrengst hadden (verlies/export). Waterretentie was alleen positief in trajecten met vijvers of reservoirs, vanwege hun wateropslagfunctie in het teeltsysteem. Er was een sterk verband tussen netto-opbrengst en rivierafvoer, hetgeen bleek uit het feit dat 93%, 60% en 67% van de opbrengsten van TSS, TP en TN werden getransporteerd gedurende 115 regendagen. Tijdens perioden met lage afvoer hadden alle landgebruikstypen positieve TSS, TP en TN opbrengsten (d.w.z. retentie), maar bij hoge afvoer waren de opbrengsten negatief (d.w.z. verlies/export). Significante effecten van het landgebruik op de hellingen op retentie van sediment en nutriënten werden ook vastgesteld.

Uit de studie met experimentele plots bleek dat het watergebruik van visvijvers (gemiddeld 101 mm m^{-2} d^{-1}) ongeveer de helft was vergeleken met rijstteelt (191 mm m^{-2} d^{-1}). De biomassatoename in visvijvers was veel lager (30 g m^{-2} in 8 maanden) dan in rijst (2500 - 4500 g m^{-2} in 4 maanden) en wetland plots (1300- 1600 g m^{-2} in 8-11 maanden). Gedurende de twee experimentele seizoenen hielden hogere concentraties van TSS, TP en TN in de toevoer- en afvoerkanalen vooral verband met menselijke activiteiten (schoonmaken van kanalen, ploegen, onkruidbestrijding en bemesting, en aflaten en dreggen van de vijvers). Vijvers en rijstvelden hadden consistent hogere TSS in hun afvoerkanalen (5-9506 mg L^{-1} voor visvijvers en 7-2088 mg L^{-1} voor rijstvelden) dan in hun toevoerkanalen (7-120 mg L^{-1} voor visvijvers en 9-483 mg L^{-1} voor rijstvelden). Voor TN en TP waren de resultaten minder duidelijk, maar ze vertoonden pieken tijdens perioden waarin het land geploegd werd, tijdens onkruid wieden en bemesting of tijdens het aflaten en dreggen van visvijvers. In wetland plots waren TSS en TN concentraties significant lager in de afvoerkanalen dan in de aanvoerkanalen, wat verband hield met de afwezigheid van verstoringen en, waarschijnlijk, door andere processen in het wetland (hogere neerslag van sediment, adsorptie en opname van nutriënten en denitrificatie).

Ploegen en onkruid wieden gedurende de eerste drie maanden van de rijstteelt, en waterverversing en dreggen in het midden en aan het einde van de visteeltperiode leiden tot de productie en lozing van grote hoeveelheden sediment, stikstof en fosfor in de afvoer van deze landgebruikssystemen. Hierdoor functioneren visvijvers als een tijdelijke opslag van sediment en nutriënten aan het begin van de teeltperiode, maar als een bron van deze stoffen aan het eind van de teeltperiode. Dit in tegenstelling tot de rijstteelt, die sediment en nutriënten genereert in het begin van de teeltperiode (bij het ploegen, onkruid wieden, overplanten en de bemesting) maar ze juist vastlegt aan het eind van de teelt. Ondanks de hogere bemesting in de rijstteelt nam de hoeveel stikstof en fosfor in de bodem af in de rijstteelt (met respectievelijk 4.7 en 1.4%) terwijl die bij visvijvers (respectievelijk 3.3% en 4.4%) en wetlands (3.8% en 1.0%) juist toenam. The afname in de bodem werd veroorzaakt door hogere opname van stikstof en fosfor in rijst (gemiddeld 662 and 270 mg m^{-2} d^{-1} respectievelijk voor N en P) dan in wetlands (359 en 121 mg m^{-2} d^{-1} N and P) en visvijvers (7.4 en 4.4 mg m^{-2} d^{-1} N and P). Voor het verbeteren van de bestaans-mogelijkheden en de economische ontwikkeling zonder negatieve effecten op de waterkwaliteit benedenstrooms van de landbouwgebieden is het belangrijk om water, sediment en nutriënten efficiënt te gebruiken en hergebruiken, bijvoorbeeld door gewasrotatie met gebruik van sediment uit visvijvers. Ook zouden wetlands kunnen worden geïntegreerd met rijst- en visteelt om als buffers voor sediment en nutriënten te kunnen dienen tijdens de kritieke perioden (ploegen en onkruid wieden, aflaten van vijvers). Zulke technieken kunnen een bijdrage leveren aan duurzame landbouw met minder erosie en lagere sediment- en nutriëntenafvoer naar benedenstroomse gebieden.

Contents

1. General introduction

1.1 Wetland ecosystem services

Agriculture is vital for economic growth and food security, but also has an impact on ecosystem services and benefits that can be lost when wetlands are converted to farming land. It is estimated that worldwide, around 35% of the wetlands were lost in the period 1970-2015, while rice farms and water reservoirs doubled in that period. With this loss rate, wetlands disappear three times faster than forests (Ramsar Convention on Wetlands, 2018). While rice paddies and reservoirs increased food and energy production, wetland ecosystem services like regulation of hydrology, water quality, climate and biodiversity were threatened (Ramsar Convention on Wetlands, 2018). Wetland conversion/drainage destroys wildlife habitat, weakens the storage capacity of wetlands for water, sediment and nutrients and, hence, increases the risks of floods, droughts, sedimentation and eutrophication of water bodies and emission of greenhouse gases (Moomaw et $al.$, 2018; Ramsar Convention on Wetlands, 2018).

It is expected that wetlands can play a significant role in achieving sustainable development and reversing the global crisis of water pollution and climate change (The Ramsar Convention on Wetlands, 2018). The Ramsar Convention on Wetlands (2018) stresses that wetlands contribute to 75 indicators of the UN Sustainable Development Goals (SDGs), while The Millennium Ecosystem Assessment (MEA, 2005) identified wetlands as strong climate change buffering systems, as they are the world largest sink of carbon. Wetlands are also potential regulators of climate change (Reddy and Delaune, 2008), but their contribution to global warming may be positive or negative, particularly on a short term basis (ASWM, 2005). Depending on hydrological and biogeochemical processes, wetlands can function as net sequesters or producers of greenhouse gases such as CO_2, CH_4 and N_2O (Pant et $al.$, 2003). Carbon is sequestered through high rates of carbon dioxide removed from atmosphere through photosynthesis and stored into wetlands through reduced rates of organic matter decomposition (Pant et $al.$, 2003). A certain portion of carbon (20%) is lost through CH_4 atmospheric emissions (Kadlec and Wallace, 2009). On the other hand, wetland drainage, peat extraction, cultivation and other disturbances lead to change in hydrology and accelerated decomposition of stored organic matter into greenhouse gases (CO_2, CH_4, N_2O). Wetlands then lose their carbon sink function and become an important source of carbon dioxide, methane gas and nitrous oxide, contributing to global warming (Moomaw et $al.$, 2018). Changes in wetland use, such as increased drainage and high nitrogen loading rates, result in increased emissions of CO_2 and N_2O (Reddy and Delaune, 2008). Methane gas is also recognized as an important greenhouse gas emitted from natural wetlands, tropical and subtropical irrigated rice (Reddy and Delaune, 2008; ATTRA, 2009). Kadlec and Wallace (2009) argued that 2.2% of wetland nitrogen loss relates to N_2O. CH_4 and N_2O emissions in rice farms are highly dependent on farming practices (irrigation schedule and fertilizer application rates). Zucong et $al.$ (1997) reported that intermittent irrigation in rice farms, alternating anaerobic-aerobic conditions stimulates emissions of N_2O while depressing CH_4 emissions, compared with permanently flooded farms that increase CH_4 emissions while lowering N_2O emissions. It was also reported that higher fertilizer application stimulates N_2O emissions while depressing the CH_4 emissions (Zucong et al., 1997; Pang et al.2009). The conversion of rice farms to fishponds is associated with a significant reduction of CH_4 and N_2O emissions, estimated as 48% and 56% by (Hu et al., 2015), and owing to lower availability of carbon substrates that can mediate CH_4 and N_2O production in ponds than in rice farms.

Similar to carbon, wetlands can store large amount of nitrogen and phosphorus in soil, sediment and biomass and release them when disturbed. The mechanisms involved in nitrogen and phosphorus storage include settling, adsorption, precipitation, and plant uptake (Denny, 1997; Verhoeven et al., 2006; Reddy and Delaune, 2008; Kadlec and Wallace, 2009). This is more effective in the tropics than in temperate zones where retention rates can mitigate the non-point nutrient sources from agriculture (Verhoeven et al., 2006). However, if overloaded with pollutants, wetlands shift from being net sequesters to sources of oversaturated pollutants (Mitsch et al., 2005; Verhoeven et al., 2006). Sustainable use of wetlands (or "wise use" as it is also called; Ramsar Convention, 2005) aims at the use of wetlands for food production while at the same time enhancing wetland retention of carbon, nitrogen and phosphorus to reduce greenhouse emission and eutrophication of water bodies.

1.2 Rehabilitation and restoration of wetland ecosystem services

Following the worldwide wetland decline and associated losses of wetland ecosystem services, it is very important to take action for conservation of the remaining wetlands and rehabilitation or restoration of wetlands that are already degraded. This is captured in the Ramsar Convention Strategic Plan, targets 5, 9 and 12 relating to maintaining ecological character, integrated resource management and ecosystem restoration, respectively and in the Sustainable Development Goals (SDGs) 6.6 and 15.1, related to the protection/restoration of wetlands and conservation of freshwater ecosystems, respectively (Ramsar Convention on Wetlands, 2018). Restoration exercises can learn from past successful cases, and scientific investigations are needed to identify options for successful restoration (Henry and Amoros, 1995). This avoids uncertainty about what combination of characteristics and processes leads to the establishment of desired ecosystem functions (Mitsch et al., 1998), the corresponding ecosystem services and values, and the management required (Ohio EPA, 2004).

Wetland restoration activities should be planned in such a way that further degradation is prevented, re-establishing the wetland integrity in structure, composition, hydrological, physicochemical and biological processes, and making the ecological functions healthy and resilient (USEPA, 2000). Integrated landscape management with appropriate combination of natural wetlands and farming lands can increase the diversification and co-existence of provisioning and ecological services. Verhoeven et al. (2006) proposed a proportion of 2-7% of wetland vegetation in riparian zones to significantly improve water quality in the catchment. Small ponds have also played a big role in regulating the water quality. A very inspiring case is the ancient multi-pond system that used to collect non-point source

2

agricultural nutrient wastewater and feed it to rice and other crops in Liuchahe catchment (China). The system was able to reach more than 90% of nutrient removal through removal mechanisms such as sedimentation, adsorption, recycling through irrigation and uptake by aquatic plants (Yan *et al.*, 1998). Other cases are: the old fishpond landscapes that existed in the Czech Republic before the 16th century, which balanced ecological and provisioning services including regulation of hydrology, water quality and biodiversity and provision of water, food and cultural needs to people (Pokorný and Květ, 2016); and the "wetland-based integrated aquaculture–agriculture systems" known as Fingerponds system which was investigated in Kenya, Uganda and Tanzania (2002-2006). Although the Fingerponds system remained at experimental scale, the adoption of such a system could contribute to increasing the provisioning services while maintaining the natural hydrology of and associated wetland ecosystem services (Kipkemboi, 2006). These cases provide a good inspiration for the rehabilitation of wetland ecosystem functions in a land-stressed environment like the Migina Catchment in southern Rwanda through integration of natural wetlands, ponds and farming to balance livelihoods of the local people while maintaining the ecosystem services.

1.3 Significance of the study

This study assesses the effects of conversion of natural wetlands to agricultural farms on the water quality and retention of sediments, nitrogen and phosphorus in valley bottoms of the Migina catchment in Southern Rwanda. It also assesses land use options that can enhance the water quality and retention of sediment and nutrients in the catchment. This is important, not only for Rwanda, because many wetlands have been and are still being converted to farmland throughout eastern, western and southern Africa (Rebelo *et al.*, 2010; van Asselen *et al.*, 2013). While these developments are positive for food security, their environmental impacts may be negative because of the alteration or destruction of wetland ecosystems and biodiversity, and the associated impact on regulating ecosystem services. Not much is known in detail about the effects of conversion of wetlands into farming on water and nutrient pathways and cycling processes. It is important to understand the effect of wetland conversion on sediment and nutrient flows to achieve a balance between the benefits of farming and maintaining the ecosystem services provided by natural wetlands. This study will increase the knowledge about water quality dynamics including sediment, nitrogen and phosphorus processes in relation to seasonal variation of rainfall, water discharge, and land use types and practices. The Migina catchment is used as a model, but findings can be applicable to other catchments with similar drivers of change, in Rwanda and in surrounding African countries.

Due to the hilly topography, land cover clearance and over-exploitation of land, Rwanda has been losing soil and nutrients (NPK) at dramatic rates. In 1980, the country soil loss was estimated at 10.1 t ha^{-1} y^{-1} (World Bank, 2005). It is unlikely that the 1980s soil loss rates have decreased because population pressure has kept increasing without appropriate soil conservation measures. Nutrient loss in central and east African countries is estimated to average more than 60 kg NPK kg ha^{-1} y^{-1} (Stoorvogel and Smaling 1990, Unfinished Agenda

1991). Rwanda's nutrient losses through soil erosion are higher than any neighbouring country, with nutrient loss rates estimates between 120 and 136 kg NPK kg ha^{-1} y^{-1} (Henao and Baanante, 1999; Unfinished Agenda, 1991). As a result, the country annually loses the capacity to feed 40,000 persons, equivalent to 1.9 % of national GDP (NISR, 2008).

Soil and nutrient loss also result in water quality problems downstream, made worse by nutrient pollutants from other sources. Most of the Rwandan towns are built on hills and mountains. Wastes are discharged downstream into wetlands and combined with nutrient pollutants from diffuse sources such as fertilized farms and urban areas. Access to adequate basic sanitation in 2005 was only 8% in rural areas and 10% in urban areas (AfDB/OECD, 2007). Failure of wetlands in upper catchments to retain and recycle nutrients and sediments (erosion) has led to downstream eutrophication and sedimentation of river and lake beds in Rwanda and in the wider Lake Victoria basin (Cheruiyot and Muhandiki, 2014). There is an urgency to find a way to retain and recycle sediment and nutrients within the catchment to maintain the catchment soil value and avoid pollution downstream.

1.4 Hypotheses of the study

As water flows through landscapes with different land use (e.g. fishponds, rice fields and wetlands), nutrient cycling occurs through the interactions among nutrients and soil, sediments, microorganisms, litter, plants, atmosphere and the water. The extent to which water quality is affected, and sediment, nitrogen and phosphorus are retained or recycled depends on physical forces like the effects of hydrology, wind, vegetation, animals and humans, and in-situ biochemical processes like photosynthesis, biomass growth, death and decomposition. By these processes, sediments, nitrogen and phosphorus can accumulate in sediment, groundwater or biomass, be lost to the atmosphere (for nitrogen), exported through harvesting or lost to downstream areas through the effluent. Land use types with low flow velocities and dense vegetation cover favour sediment settling more than land uses with fast and turbulent flows and low vegetation density. Therefore, fishponds and natural wetlands are expected to contribute more to nutrient and sediment retention than rice fields. Increasing their use in valley bottoms can increase the overall retention function of valley bottoms.

Fishponds, rice farms and "natural" wetlands may be integrated for optimizing valley-bottom sediment and nutrient retention efficiency, e.g., sediments and nutrients from the feeding river can settle in the fishpond. Accumulated sediments and nutrients can be re-used for rice cultivation in the same pond or be removed (by dredging) to be re-used for cultivation of other crops. Sediments from fishponds and rice farms can be trapped in a natural wetland that provides enhanced sediment retention capacity through the low flow velocity, the presence of sediment settling plain, and the dense vegetation and litter.

Nutrient retention and recycling also depend on groundwater exchanges, nutrient concentrations in the inflow, sorption and desorption, pH and redox potential, and microbial community. For this reason it is important to look at whole nutrient cycles and to consider combinations of land use types for optimal nutrient and sediment cycling and retention.

1.5 Research objectives

This research aimed at assessing the effects of conversion of natural wetlands to agricultural farms on the water quality and retention of sediments, nitrogen and phosphorus and to identify land use/land cover characteristic and management options that can enhance water quality and sediment and nutrient retention capacity. One central research question is whether these wetland types retain nutrient and sediments from disturbed uplands, which are dominated by urban and agricultural uses. Migina catchment in Rwanda was selected as a study model. The research approach used different landscape-scale synoptic surveys (catchment, reaches and experimental plots) of land use and water quality variables. The specific objectives are:

1. To characterize water quality in Migina catchment in relation to river discharge and land use and land cover (LULC);
2. To relate trends in the retention of total suspended solids (TSS), total phosphorus (TP) and total nitrogen (TN) to valley bottom land use and land cover (LULC), in the context of seasonal rainfall and landscape features;
3. Using experimental mesocosms with different LULC to assess the effects of conversion of wetlands to fish and rice farming on water quality; and
4. Using experimental mesocosm with different LULC to assess the effects of conversion of natural wetlands to rice and fish farming on the pathways of sediment, nitrogen and phosphorus.

This thesis is composed of six chapters. Chapter 1 (General introduction) discusses the importance of wetlands in regulating the hydrology, water quality, climate change. It also discusses environmental problems associated with wetland conversion to increase the provisioning services. From this, the chapter presents the rationale for this study, the study hypotheses and defines the objectives of the study.

Chapter 2 discusses the effects of river discharge and land use and land cover (LULC) on the water quality in the Migina catchment. The study shows the relationship between water quality and discharge for different LULC at the catchment scale (Munyazi, Mukura and Akagera sub-catchments) and at the reach scale (16 reaches of Munyazi sub-catchment).

Chapter 3 investigates the effects of agricultural LULC on nutrient and sediment retention in valley bottom wetlands. The study uses 16 Munyazi sub-catchment reaches with different LULC (ploughed, rice farming, vegetable farming, pond/dam and grass/forest) and landscape features (slope, length, area, population density), rainfall and water discharge to assess the seasonal variation in the retention of total suspended solids (TSS), total phosphorus (TP) and total nitrogen (TN).

Chapter 4 investigates the effects of conversion of wetlands to fish and rice farming on water quality. The study uses 6 replicates for each of three mesocosms with different LULC (rice farms, fishponds and wetlands) at the Rwasave fishpond station to assess the water quality dynamics from the inflow to the outflow, in relation to hydrological flows, biomass growth and land use practices (land works and feed or fertilizer application).

Chapter 5 evaluates the effects of conversion of wetlands to fish and rice farming on sediment, nitrogen and phosphorus pathways. The study uses 6 replicates for each of the three mesocosm LULC types (rice farms, fishponds and wetlands) at the Rwasave fishpond station to assess sediment, nitrogen and phosphorus pathways across the study LULC.

Chapter 6 gives a synthesis of the key study findings and gives conclusions and recommendations.

2. Effects of river discharge and land use and land cover (LULC) on water quality dynamics in Migina catchment, Rwanda[1]

Abstract

Agricultural intensification may accelerate the loss of wetlands, increasing the concentrations of nutrients and sediments in downstream water bodies. The objective of this study was to assess the effects of land use and land cover (LULC) and river discharge on water quality in the Migina catchment, southern Rwanda. Rainfall, discharge and water quality (total nitrogen (TN), total phosphorus (TP), total suspended solids (TSS), dissolved oxygen (DO), conductivity (EC), pH, and temperature) were measured in different periods from May 2009 to June 2013. In 2011, measurements were done at the outlets of 3 sub-catchments (Munyazi, Mukura and Akagera). Between May 2012 and May 2013 the measurements were done in 16 reaches of Munyazi dominated by rice, vegetables, grass/forest or ponds/reservoirs. Water quality was also measured during two rainfall events. Results showed seasonal trends in water quality associated with high water flows and farming activities. Across all sites, TSS related positively to discharge, increasing 2-8 times during high flow periods. EC, temperature, DO and pH decreased with increasing discharge, while TN and TP did not show a clear pattern. TSS concentrations were consistently higher downstream of reaches dominated by rice and vegetable farming. For TN and TP results were mixed, but suggesting higher concentration of TN and TP during the dry and early rainy (and farming) season, and then wash out during the rainy season, with subsequent dilution at the end of the rains. Rice and vegetable farming generate the transport of sediment as opposed to ponds/reservoir and grass/forest.

Keywords: Agriculture, discharge, land use, nutrients, water quality, wetlands

---

[1] Published as:
Uwimana, A., van Dam, A.A., Gettel, G.M., Bigirimana, B., Irvine, K., 2017. Effects of river discharge and land use and land cover (LULC) on water quality dynamics in Migina Catchment, Rwanda. Environmental Management 60, 496 - 512. https://doi.org/10.1007/s00267-017-0891-7

2.1 Introduction

Pollution by nutrients is recognized as the most widespread cause of water quality degradation (UN-WWAP, 2009), with runoff from agriculture a major source of nutrient and sediment pollution (USEPA, 2005). In sub-Saharan Africa, agricultural intensification is critical to increase food security and economic development. Intensification occurs both by increasing the area of land under cultivation, and through an increase of fertilizer inputs and irrigation. Despite international recognition of the importance of wetlands and national policies for their protection, contradictory policies and practice lead to the conversion of wetlands to farmland throughout eastern, western and southern Africa. Wetland conversion also occurs informally, with wetlands used in the dry season for production of sugarcane, groundnut, vegetables and fruits, as well as for grazing livestock (McCartney et al., 2010; Schuyt, 2005; Wood et al., 2013). Loss of wetlands, often involving increases in fertilizer and irrigation, enhances risks of increased erosion and nutrient export (Verhoeven et al., 2006). Despite these pressures, there are very few field studies of the effects of agricultural land use on runoff and water quality in sub-Saharan Africa.

In northern hemispheres, many wetland ecosystems have been shown to regulate water quality through ecological and biogeochemical processes, functioning as sinks for sediment, nutrients and pollutants (e.g. Johnston, 1991; Zedler, 2003). Verhoeven et al. (2006), reviewing some examples from the USA, Sweden, and China, concluded that wetlands may contribute significantly to water quality improvement if they constitute 2-7% of the surface area of the catchment. Water quality improvement occurs through vegetation, biogeochemical processes, and increased water residence time which can control sediment and nutrients, and this may be especially important when small reservoirs created for aquaculture and irrigation are constructed (Rădoane and Rădoane, 2005; Reddy and Delaune, 2008; Weissteiner et al., 2013). Human activities like deforestation (Milliman and Syvitski, 1992; Mkanda, 2002) and farming (Hecky et al., 2003) can lead to elevated losses of sediment and nutrients. Wetland ecosystems in Sub-Saharan Africa comprise a similar component of the landscape, with an estimated 65% of African wetlands occurring in the four major river basins (Chad, Congo, Niger and Nile), which are rapidly degrading (McCartney et al., 2010; Davidson, 2014). Only a few studies are available on the effect of land use on water quality in African catchments (Bagalwa, 2006; de Villiers and Thiart, 2007; Dunne, 1979; Hecky et al., 2003). These river basin scale studies are, however, based on a crude classification of land use as agriculture or forested area and do not take into account the effects of land cover within the agricultural land use. For example, it is assumed that rice, vegetable or fish farming all influence water quality in the same way. The relationship between land use, hydrology and water quality is dynamic in time and space and no single relationship has universal validity (Bartley and Speirs, 2010; Hilderbrand et al., 2005). There is a need for a more detailed understanding of the effects of land use change on water quality, particularly in African catchments.

Sub-Saharan Africa in the near future is expected to experience stronger variations in hydrology owing to climate change causing increasing variability in duration and timing of wet and dry seasons (Kotir, 2011). Depending on the type of land use, flow paths, and

8

biogeochemical processes, the relationships of sediment and nutrient concentrations with river discharge vary (Whitehead *et al.*, 2009). For example, concentration can increase with discharge when relatively small rainfall events flush accumulated nutrients in soil pore-water into streams (Bae, 2013; Carroll *et al.*, 2007; Coulliette and Noble, 2008) or when sediments are resuspended (Römkens *et al.*, 2002). The effects of land use change and wetland conversion on stream water quality therefore need to be studied in the context of hydrology and climate.

Here we use a model catchment in southern Rwanda to examine the relationship between water quality, land use and water discharge. Rwanda, which has the highest population density in Africa (490 inh./km^2; World Bank, 2015), experiences rapid conversion of wetlands to agricultural land, with government policies to convert up to 80% of wetlands, expanding the farmed area four-fold between 2006 and 2020 (MINAGRI, 2009; 2010a). The wetlands are used mostly for rice production, but also for other crops such as sugarcane, maize, beans, vegetables and subsistence-level aquaculture (MINAGRI, 2009; 2010a). The Migina catchment is one of the uppermost catchments of the Akagera river, the largest tributary of Lake Victoria in the Nile Basin, which drains 67% of the territory of Rwanda (MINITERE, 2004). The catchment is representative of the environmental problems encountered in many similar headwater catchments in the region. Overexploitation of resources, drainage of wetlands for land reclamation, settlement and urbanization, industrial development and road construction have caused water quality and wetland degradation (e.g. Nhapi *et al.*, 2011; Usanzineza *et al.*, 2011). The overall objective of this study was to characterize water quality in Migina catchment in relation to river discharge and land use. We first compared rainfall, discharge and water quality at the outlets of three sub-catchments (Munyazi, Mukura and Akagera). Then we explored the relationship between discharge and water quality during two rainfall events at the outlet of Munyazi sub-catchment. Finally, we monitored discharge and water quality throughout one year in 16 reaches of Munyazi sub-catchment with different land use and land cover (LULC). A subsequent paper (Uwimana *et al.*, in prep.) presents data on nutrient and sediment loads and retention in more detail in relation to land use and seasonal land use change in individual study reaches of Migina catchment.

2.2 Material and Methods

2.2.1 Study area

Migina catchment (Figure 2-1) in the Southern Province of Rwanda is shared between Huye, Gisagara and Nyaruguru Districts (29°42'- 29°48' E; 2°32'- 2°48' S). It is one of the uppermost catchments of Akagera river, located between 1400 and 2247 m altitude. The landscape is mountainous with small valleys containing wetlands, streams or rivers (Figure 2-1a).

9

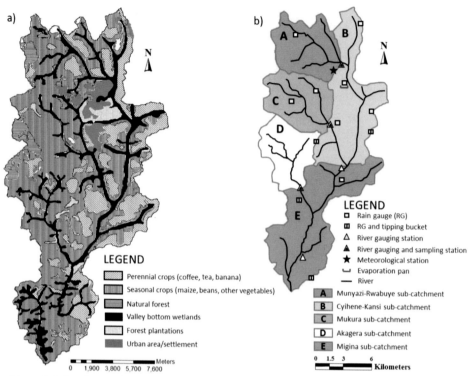

Figure 2-1. Land use and land cover (LULC) in Migina catchment, southern Rwanda (a); and sub-catchments of the Migina catchment (b). Figure 2-1b was adapted from the map in Munyaneza et al. (2012).

The catchment can be divided into five sub-catchments: Munyazi, Mukura, Akagera, Cyihene and Migina (Figure 2-1b; Munyaneza *et al.*, 2012). Agriculture dominates land use in both the valley bottoms and hill slopes (Table 2-1).

The upstream of the catchment (Munyazi, Mukura and Cyihene) is dominated by urban areas and settlements on the hillsides, and by rice cultivation in the valleys. Mukura has also a water treatment plant (Kadahokwa) that abstracts almost 6000 m^3 d^{-1} from the river. The downstream part (Akagera and Migina) is more rural and rice is less common. In rural areas, people use springs for their daily water needs. Wetland conversion into agricultural farms has become the preferred option for increasing the food security under the Strategic Plans for the Transformation of Agriculture (PSTA) and the Rwanda Irrigation Master Plan (MINAGRI, 2009; 2010a; 2013a,b). Consequently, natural wetland vegetation (often dominated by *Cyperus latifolia*) has almost disappeared throughout the catchment.

Table 2-1. Characteristics of different sub-catchments of Migina (adapted from Munyaneza 2014).

Sub-catch-ment	Area (km²)	Altitude at basin outlet (m)	Basin slopes (%)	Rainfall (mm a⁻¹)	Dominant valley bottom land use	Catchment land use (%)			
						Agri-culture	Forest	Grass-land	Urban
Munyazi	55.0	1662	15.8	1453	Rice	90	8	0	1.6
Mukura	41.6	1618	19.5	1666	Vegetables, rice	85	12	1.4	2.2
Cyihene	71.1	1577	12.5	1457	Rice	89	5.8	0	4.8
Akagera	32.2	1575	20.8	1507	Vegetables	88	12	0	0
Migina	61.1	1520	18.6	1415	Vegetables	100	0	0	0

Farming activities can be grouped into two or three seasons according the types of crops. Rice farming has two seasons (February-July and August-January), while vegetable farming has three (September-January, February-May and June-August). While fertilizer application to vegetables is variable, rice farming involves high rates of mineral fertilizer (4 kg of NPK and 2 kg of urea per acre) and manure (200 kg per acre) application. Fertilizer and manure are applied in the beginning (February-March and August-September), while urea is applied in the middle (April-May and October-November) of the rice farming season. The national average fertilizer application has increased steadily, from an average of about 4 kg ha⁻¹ in 2006 to 30 kg ha⁻¹ in 2013, and is expected to reach 45 kg ha⁻¹ in 2017/18 (MINAGRI, 2014). Land cover changes within the crop growing season can fluctuate from bare soil resulting from hoeing to a dense canopy at the end of the season.

The climate is temperate tropical humid with dry periods in January-February and June-August. Peak rainfall occurs in November and April. Average temperature is 20°C and annual rainfall around 1200 mm. The climate in the catchment is influenced by the proximity to the equator at 2° to the North, by Lake Victoria in the East, and by the altitude (1435-2247 m). The geology of the area is typical granites, quartzites and schists. Weathering and ferralization of these rocks has resulted in Ferrallitic soil with clay deposits in the valley bottoms. The geology of all sub-catchments is similar (Ministere des Ressources Naturelles et Service Geologique, 1963).

2.2.2 Field sampling design

The study was conducted in 2011 and 2012-2013. In 2011, rainfall, discharge, and water quality were monitored at the outlets of three sub-catchments: Munyazi, Mukura and Akagera (Figure 2-1b). These sub-catchments vary in dominant land cover, with rice most prominent in the Munyazi sub-catchment and vegetable farming dominating in Mukura and Akagera sub-catchments (Table 2-1). Water quality variables measured were: temperature, electrical conductivity (EC), pH, dissolved oxygen concentration (DO), total suspended solids (TSS), and total phosphorus and nitrogen concentrations (TP and TN). In May 2012 - May 2013, the study focused on 16 river reaches in the Munyazi sub-catchment to compare the effects of LULC more specifically on better isolated agricultural land cover types. During this period, water quality was measured upstream and downstream of individual reaches and LULC in each reach was observed for that period. In addition, two rainfall events were sampled opportunistically at the outlet of Munyazi sub-catchment to better understand the effects of high water flow on water quality.

2.2.3 Data collection

2.2.3.1 Catchment delineation and hydrology

A Digital Elevation Model (DEM) with land use information was provided by the University of Rwanda Centre for Geographic Information Systems & Remote Sensing (UR-CGIS) to delineate the Migina catchment, identify the hydrological network, river reaches and land use (Figures 2-1a and 2-2). Rainfall and discharge data were collected from May 2009 to December 2011 from rain and river gauging stations (Figure 2-1b). Rainfall in Munyazi, Mukura and Akagera sub-catchments was collected using manual rain gauges established at different schools within different sub-catchments and was calculated for each sub-catchment as the arithmetic average of rainfall as follows: Rwasave, CGIS, Sovu and Save stations for Munyazi; Rango, Mpare and Vumbi for Mukura; and Murama and Mubumbano for Akagera. Discharge at the outlet of each sub-catchment was calculated based on rating curves developed by Munyaneza (2014). Briefly, water pressure was measured every 30 minutes (Mini-Diver DI501, Schlumberger Water Services, Delft, The Netherlands) and used in regression analysis with monthly manual discharge measurements (Universal Current Meter, OTT C31, Hydromet GmbH, Kempten, Germany) from May 2009 – June 2011.

2.2.3.2 Water quality in three sub-catchments (2011)

Water samples were collected monthly from January to December 2011 at the outlets of the three sub-catchments (Figure 2-1) at different depths (surface, middle and bottom) using a standard 2 litre water sampler (model no. 436132; Hydro-Bios, Kiel-Holtenau, Germany) and mixed to form a composite sample of the site (Ruttner-Kolisko, 1977). A multimeter (18.50.SA Eijkelkamp, Giesbeek, The Netherlands) was used to measure EC (μS cm^{-1}), pH and temperature (°C) directly on site. EC was temperature corrected (specific conductance) at 25 °C with a temperature correction coefficient of 0.0191 as directed in Standard Methods for the Examination of Water and Wastewater (APHA, 2005). DO was measured using a Portable Dissolved Oxygen Meter Accumet Waterproof AP74 (Fisher Scientific, Pittsburgh, USA). TSS was determined on-site using a Portable colorimeter DR/890 (HACH, Colorado, USA). Samples for TP and TN were collected unfiltered and preserved by acidification to pH = 1-2, using 0.1 N sulfuric acid. Both were analysed at the University of Rwanda Huye campus water quality laboratory using the persulfate digestion method (APHA, 2005).

2.2.3.3 Discharge, water quality and LULC in 16 study reaches (May 2012 - May 2013)

To identify and classify LULC, surveys conducted in the 16 reaches using a combination of visual estimates along a transect and GPS measurements were imported into Google Maps to estimate the area of each reach. LULC was classified into four groups (rice farming, vegetable farming, grass/forest and ponds/reservoir) according to the predominance of each LULC type.

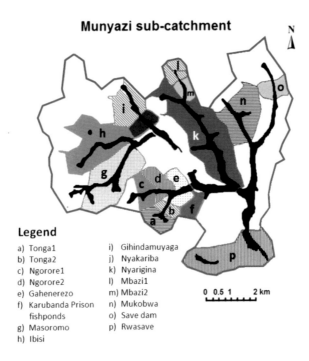

Munyazi sub-catchment

Legend

a) Tonga1
b) Tonga2
c) Ngorore1
d) Ngorore2
e) Gahenerezo
f) Karubanda Prison fishponds
g) Masoromo
h) Ibisi
i) Gihindamuyaga
j) Nyakariba
k) Nyarigina
l) Mbazi1
m) Mbazi2
n) Mukobwa
o) Save dam
p) Rwasave

0 0.5 1   2 km

Figure 2-2. The 16 reaches of the upper Migina river in the Munyazi sub-catchment, southern Rwanda used in this study. See Table 2 for land use & land cover (LULC) characteristics of these reaches.

Monthly measurements for water discharge and water quality were taken from May 2012 to May 2013 in 16 river reaches in Munyazi sub-catchment. Geographic coordinates were analysed using ESRI®ArcGIS 9, ArcMap Version 9.3.1 to delineate the 16 reaches (Figure 2-2). The monthly discharge and water measurements in the upstream and downstream sites for each reach compared inputs and outputs. Total monthly discharge for each reach was estimated from daily water pressure records at the downstream gauging station at Rwabuye (Figure 2-1b) using regression equations between measured discharge in each reach and the corresponding water pressure from the gauging station.

Water quality was also measured at the outlet of the Munyazi sub-catchment during two storm events that started during field sampling on 23 March 2013 and 5 April 2013. Most measurements were taken at 0.5-2 hour intervals from the start of the event until the peak discharge. Onset of darkness necessitated stopping sampling before the storms were finished because of security concerns. Consequently, the falling limb of the hydrograph was not sampled.

2.2.3.4 Data analysis

Regression analysis was used to relate water quality to discharge for monthly data in the 16 study reaches and for the hourly data collected from the two storm events. Model fit (evaluated using the mean of the absolute values of the residual errors of the estimates) showed that linear regression gave the best results generally. For the 16 reaches, differences between upstream and downstream water quality parameters for each month and for each reach were calculated, negative values indicating a higher downstream than upstream value. This was done for TSS, TN, TP, temperature, pH, EC and DO. The differences were analysed using analysis of covariance (ANCOVA), with LULC category (rice, vegetables, grass/forest, or ponds/reservoir) as the main categorical explanatory variables and total monthly discharge Q (natural log-transformed) as the continuous covariate. Because interactions between LULC and ln(Q) in the ANCOVA were significant in many cases, indicating different slopes in the relationships between water quality change and Q for different LULC categories, mean water quality upstream-downstream differences were also compared among LULC categories using one-way ANOVA and simple regression of water quality differences with ln(Q). R version 3.1.1 (http://www.R-project.org/) was used for all data analyses. Tests for significance were done at $\alpha = 0.05$ unless stated otherwise.

2.3. Results

2.3.1 Rainfall, discharge and water quality in three Migina sub-catchments

Rainfall and water discharge had similar patterns among sub-catchments and shared similar seasonality (Figure 2-3).

Rainfall ranged between 100 and 200 mm month^{-1} between October and May with a peak in April in all three sub-catchments, while lower rainfall (< 100 mm) was observed from June to September. These patterns agree with the typical seasonal distribution of rainfall in the area. Although seasonal patterns in rainfall were similar in all the sub-catchments, discharge varied by an order of magnitude, with the highest occurring in the Akagera sub-catchment and the lowest in the Mukura sub-catchment. For example, on 7 February 2010 the water discharge in Akagera was 10 m^3 sec^{-1} following rainfall of 67 mm, while for the same event, discharge in Mukura was 0.046 m^3 sec^{-1} for rainfall of 61 mm. Discharge followed the rainfall pattern with highest values occurring during the rainy periods (Figure 2-3). The highest monthly mean discharges (0.5 - 3.0 m^3 s^{-1}) occurred in May of both years and were observed in the Akagera sub-catchment. The lowest discharges (0 - 0.2 m^3 s^{-1}) were observed in Mukura sub-catchment, and intermediate in Munyazi (0 - 1.2 m^3 s^{-1}).

The three sub-catchments followed similar seasonal patterns for most of the water quality variables, but varied with respect to their relationship with discharge. Lower values of temperature, pH and EC and higher concentrations of TSS were observed in June and November during high rainfall and discharge periods, while low TSS was observed in August and September during the dry season. Relatively higher DO was observed in the downstream sub-catchment (Akagera) in the dry seasons (February and May-August). Specifically, water temperature ranged from 18 to 27 °C, with the highest temperatures occurring during the driest months and in Mukura, the sub-catchment with the lowest discharge (Figure 2-4). EC followed a similar pattern, and fluctuated around 100 µS cm^{-1} for all sub-catchments, with Mukura sub-catchment reaching the highest values up to 400 µS cm^{-1} during the dry period (Figure 2-4c). DO varied between 2.8 and 7.6 mg L^{-1} (Figure 2-4d), with the lowest values (around 3 mg L^{-1}) observed in Munyazi (upper sub-catchment) in April and May and highest values of 6-7 mg L^{-1} in all sub-catchments in June. The pH ranged from 5.7 to 7.0 and did not vary remarkably among the sub-catchments. The lowest pH was associated with the high rainfall events in June and November (Figure 2-4b).

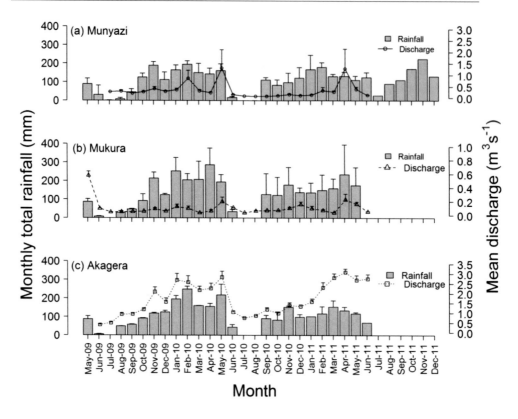

Figure 2-3. Monthly total rainfall (means of several rain gauges in mm) and mean river discharge (monthly means of daily discharge measurements in m³ s⁻¹) in three sub-catchments (Mukura, Akagera and Munyazi) of Migina catchment in the period May 2009 - June 2011 (for Munyazi rainfall until December 2011). Error bars represent standard deviation.

TSS varied widely from 6 to 457 mg L⁻¹, and maximal during high flow. Like TN and TP, highest values were recorded in the Munyazi sub-catchment, even though it generally experienced intermediate values in discharge compared with the other sub-catchments (Figure 4e). TN and TP did not follow these patterns, likely as a result of a combination of high runoff and farming seasons. TN and TP concentrations were in the range 0.02-16.5 and 0.04-1.5 mg L⁻¹, respectively. Both TN and TP showed highest concentrations during the rainy periods (2.3-16.5 mg L⁻¹ for TN and 0.2-1.5 mg L⁻¹ for TP), but peaks occurred indifferent months. Highest TP was found in April (1.2 mg L⁻¹) and November (1.5 mg L⁻¹) during the main farming season (September-January). Highest TN was recorded in June (16.5 mg L⁻¹) during a high flow event (Figure 2-4 f, g), suggesting a mobilization, rather than dilution, of nitrogen.

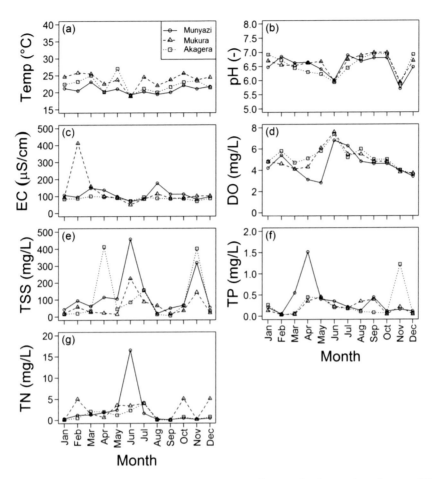

Figure 2-4. Monthly measurements of temperature (a), pH (b), specific electrical conductivity (EC, c), dissolved oxygen concentration (DO, d), total suspended solids (TSS, e) total phosphorus (TP, f), and total nitrogen (TN, g) at the outlet of three sub-catchments (Munyazi, Mukura and Akagera) of Migina river, Rwanda from January to December 2011.

With respect to the relation between water quality and discharge at the time of sampling, only pH (decreasing) and TSS (increasing) showed a clear linear relationship with discharge in this range (0 - 1.5 m^3 s^{-1}) for all three sub-catchments (Figure 2-5).

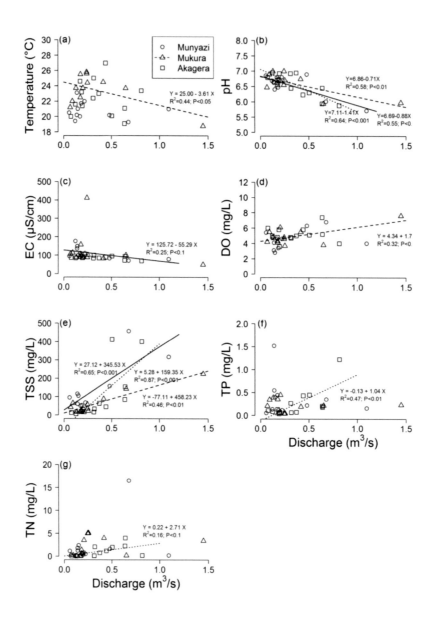

Figure 2-5. Relationships between discharge and water quality parameters in three sub-catchments (Munyazi, Mukura and Akagera) of Migina river, Rwanda between January and December 2011. Regression lines and equations are shown when significant (P < 0.05 or better) or marginally significant (P < 0.1).

2.3.2 Rainfall, discharge and water quality in Munyazi sub-catchment

Rainfall and discharge at the outlet of Munyazi in the period May 2012 - May 2013 followed similar seasonal patterns of the previous year, with a dry period from July to late August and rainfall during the months September 2012 to May 2013 (Figure 2-6). The discharge ranged from 0.2 $m^3 s^{-1}$ in July 2012 (when rainfall was zero) to 11 $m^3 s^{-1}$ in April, corresponding with 360 mm of rain that month, including one day with 75 mm of rain. In other wet months total rainfall was around 100 mm.

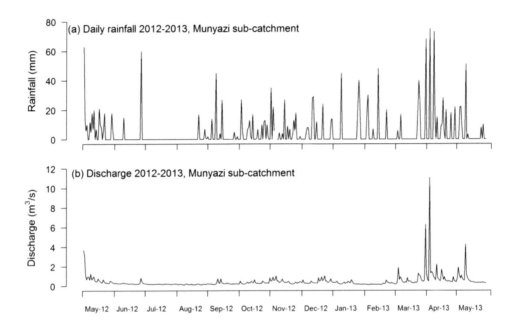

Figure 2-6. Daily rainfall (mm) and discharge ($m^3 s^{-1}$) at the outlet of Munyazi sub-catchment in the period May 2012 - May 2013.

The two rain events measured at the outlet of Munyazi were different with respect to intensity and duration, but showed similar relationships between discharge and water quality. The March 2013 event lasted the whole day with discontinuous light rain showers with peak discharge at 3.8 $m^3 s^{-1}$. In contrast, the April event was characterized by heavy and continuous rainfall that lasted about 4 hours, with flooding around the river bank. Peak discharge was much higher at 21.9 $m^3 s^{-1}$ (Figure 2-7a).

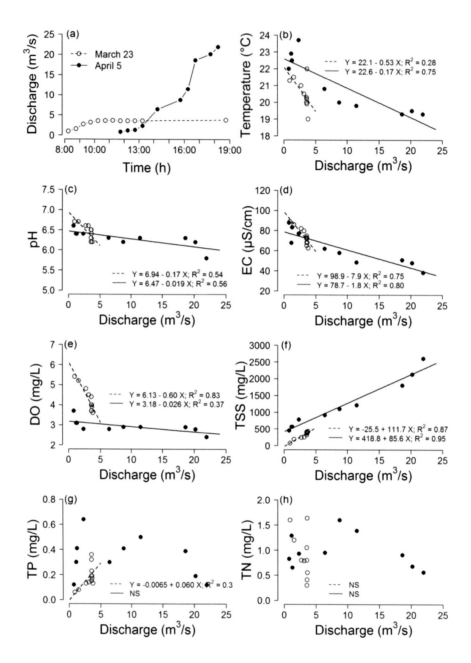

Figure 2-7 Discharge, TSS, TP and TN for two high flow events (23 March and 5 April 2013) in Migina catchment (Munyazi sub-catchment), southern Rwanda. (a) Hydrograph; (b)-(h) Relationship between discharge and water quality variables.

Despite the differences in runoff, temperature, pH, EC, and DO all decreased significantly (p < 0.05) with increasing discharge (linear regression, $R^2 = 0.28$ and 0.75 for temperature, 0.54 and 0.56 for pH, 0.84 and 0.37 for DO, 0.75 and 0.80 for EC, for March and April events respectively). Higher slopes in all of these parameters were observed when discharge increased from 0 to 5 $m^3 s^{-1}$ during the smaller March event. TSS was positively related to discharge (linear regression, $R^2 = 0.87$, p < 0.0001 for March event and $R^2 = 0.95$, p < 0.0001 for April event), increasing from less than 500 mg L^{-1} at low discharge to around 2000 mg L^{-1} at discharge higher than 20 $m^3 s^{-1}$ (Figure 2-7f). In contrast to above variables, no clear relationship was observed between TP and TN concentrations and discharge (Figure 2-7g,h). Somewhat higher TP and TN concentrations were observed at intermediate discharge (up to 15 $m^3 s^{-1}$) but a clear pattern was not visible. DO and TSS were inversely related ($R^2 = 0.96$, p <0.001 for the March event and $R^2 = 0.47$, p <0.05 for the April event; Figure 2-8).

2.3.3 Water quality as affected by season and LULC in 16 reaches of Munyazi

The 16 reaches were classified into four LULC groups (rice farming, vegetable farming, grass/forest and ponds/reservoirs) according to the dominant LULC type (Table 2-2). Some reaches were classified slightly differently as follows (see Table 2-2): Rwasave reach (30% fishponds and 46% rice farms) was classified as ponds/reservoirs because fishponds were consistently present throughout the year while rice was absent in some periods during fallow and ploughing. Similarly, Nyakariba reach (40% grass and 60% vegetables) was classified as a grass/forest reach. Two reaches were quite different than the other 14: The upstream of Ibisi (Ibisi bya Huye) was dominated by grass and forest and Karubanda had the highest human population density because of the Karubanda Boarding school and local prison.

Water temperature varied, from 16.3 to – 33.5 °C, with seasons and LULC (Figure 2-9). Highest temperatures were observed at low discharge between June 2012 and February 2013 (Figure 2-9), and were higher downstream of rice farming than vegetable farming and grass-forest (Figure 2-9a1, a2 and a3), and similar to observations in the rice-dominated sub-catchments of Munyazi and Mukura, and vegetable-dominated Akagera sub-catchment) (Table 2-1). The lowest temperatures were observed during heavy rainfall (March-May), and in downstream reaches dominated by grass/forest (Figure 2-9a3). The variation in pH of 5.1 to 8.6 (Figure 2-9b) was larger than that observed in the main sub-catchments. Minimal pH occurred during the wet season (April and May 2013) in all sub-catchments. Highest pH was observed downstream of ponds and reservoirs, especially during the dry periods, presumably because photosynthesis drove up pH values. As in the three sub-catchments, EC on the 16 Munyazi reaches fluctuated around 100 μS cm^{-1}, with an exceptional median value of 208 μS cm^{-1} in the Karubanda reach (Figure 2-9c4) which releases lower discharge probably as a result of fish farming and release of untreated wastewater from the prison. DO fluctuated widely between 0.5 and 8.7 mg L^{-1} (Figure 2-9d), with high values occurring during hot, sunny conditions (e.g. in July and November 2012) and downstream of rice farming and ponds. Low DO values (2-4 mg L^{-1}) were observed downstream of reaches with grass/forest.

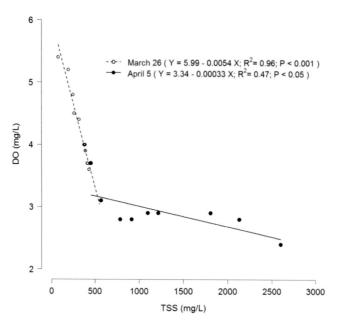

Figure 2-8. Relationships between TSS and DO for the 2 high flows of 23 March 2013 and 5 April 2013.

TSS and TN across the Munyazi reaches showed patterns similar to those in the three sub-catchments, with high concentrations occurring during high flow periods (March-May 2013) or during the main agricultural season (September-January). Despite the similar patterns, TSS was more variable than observed in the sub-catchments, from 0 to 1318 mg L^{-1}, with higher values downstream of reaches with rice and vegetables (Figure 2-9e), and lowest values in June-August, the period following harvest and characterized by low flow and recolonization of grass on fallow agricultural land. TP concentration was somewhat less variable than observed in the sub-catchments, ranging from 0 to 0.87 mg L^{-1} (Figure 2-9f) with low values in June-July, directly following the rainy period, while elevated concentrations of TP were observed during the main farming season (September to January). TN concentrations varied in the closest range to the observed in the sub-catchments, from 0.2 to 17.1 mg L^{-1} (Figure 2-9g), with low values associated with the period following the rainy period (June and July). Higher values were observed from the end of the dry season (August) until December (period of intensive farming activities).

Significant (P<0.05) differences between upstream and downstream water quality were found for some water quality variables among LULC types (ANCOVA, Table 2-3; ANOVA, Table 2-4; see also Figure 2-10).

Table 2-2. Land use and land cover (LULC) in % of reach area for 13 reaches in Migina catchment (Munyazi sub-catchment), southern Rwanda. Numbers are means of 13 months and coefficient of variation) in the period May 2012 - May 2013. See Figure 2-2 for location of reaches. LULC groups were determined on the basis of these percentages. See text for more explanation.

| LULC group | Reach | Land use & land cover (LULC, % of area) | | | |
		Rice farming	Vegetable farming	Ponds/ reservoir	Grass/ forest
Rice	Tonga-2	100 (0)	0	0	0
Rice	Nyarigina	90 (0)	10 (0)	0	0
Rice	Mukobwa	89 (0)	11 (0)	0	0
Rice	Mbazi-1	89 (0)	11 (0)	0	0
Rice	Gahenerezo	87 (0)	13 (0)	0	0
Vegetable	Masoromo	0	100 (0)	0	0
Vegetable	Ngorore-1	0	100 (0)	0	0
Vegetable	Mbazi-2	0	80 (49)	0	20 (107)
Vegetable	Ibisi bya Huye	12 (40)	73 (20)	0	15 (0)
Ponds/Reservoir	Save Dam	0	10 (0)	50 (0)	40 (0)
Ponds/Reservoir	Karubanda Fishponds	0	50 (0)	50 (0)	0
Ponds/Reservoir	Rwasave	46 (21)	9 (0)	30 (0)	15 (0)
Grass/forest	Tonga-1	0	0	0	100 (0)
Grass/forest	Gihindamuyaga	0	10 (45)	0	90 (3)
Grass/forest	Ngorore-2	0	32 (41)	0	68 (31)
Grass/forest	Nyakariba	0	60 (0)	0	40 (0)

TSS was consistently higher downstream of rice and vegetable reaches (mean differences 95 and 127 mg L^{-1}, respectively), and consistently lower downstream of grass/forest and ponds/reservoir reaches (75 and 67 mg L^{-1}, respectively; see Figure 2-10 and Table 2-3). For TN and TP, however, upstream-downstream differences were small (maximum 0.5 mg L^{-1}) and not significantly different between LULC types (Table 2-3). Temperature was significantly lower downstream of grass/forest and ponds/reservoir reaches (1.4 and 0.3 °C on average, respectively) and higher downstream of rice and vegetable reaches (1.3 and 0.4 °C, respectively). pH and DO were consistently lower downstream of grass/forest (mean differences of 0.12 pH units and 1.6 mg L^{-1} of DO), and consistently higher downstream of rice and vegetable reaches (mean difference of 0.19 and 0.22 pH units for rice and vegetable respectively, and 0.5 and 0.8 L^{-1} of DO for rice and vegetable respectively. EC was particularly higher downstream of ponds/reservoir reaches with mean difference of 46 µS cm⁻¹ (Table 2-3).

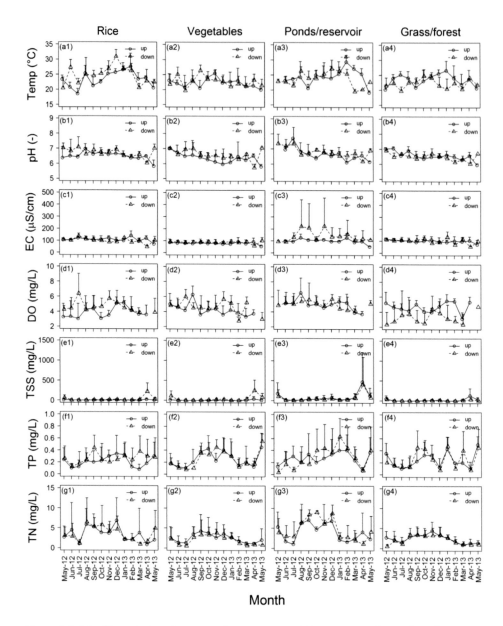

Figure 2-9. Monthly variation of water quality upstream and downstream of 16 reaches of Munyazi sub-catchment with different land use & land cover (LULC, see Table 2) in the period May 2012-May 2013. LULC include rice farming, vegetable farming, grass/forest and ponds/reservoir.

Table 2-3. Mean differences (± standard deviation) between upstream and downstream water quality parameters in four land use/land cover (LULC) types in 16 reaches of Migina catchment (Munyazi sub-catchment), southern Rwanda. Negative values indicate a higher value downstream than upstream. Means and standard deviations result from 13 months (May 2012-May 2013). Means sharing the same letter were not significantly different (one-way ANOVA, P<0.05).

LULC	Temp (°C)	pH	EC (µS cm⁻¹)	DO (mg L⁻¹)	TSS (mg L⁻¹)	TP (mg L⁻¹)	TN (mg L⁻¹)
				Average upstream-downstream difference			
Rice	-1.3 ± 4.4[b]	-0.19[a]	10.8 ± 38.1[a]	-0.8 ± 2.1[a]	-95.0 ± 122.7[a]	-0.052 ± 0.25[a]	-0.051 ± 3.23[a]
Vegetables	-0.43 ± 3.9[ab]	-0.22[a]	-8.9 ± 24.5[a]	-0.5 ± 1.8[a]	-127.0 ± 182.0[a]	-0.0019 ± 0.20[a]	0.50 ± 1.46[a]
Ponds/reservoir	0.33 ± 4.2[ab]	-0.13[ab]	-46.3 ± 105.6[b]	0.1 ± 1.5[a]	66.9 ± 147.8[b]	-0.013 ± 0.34[a]	-0.73 ± 3.81[a]
Grass/forest	1.4 ± 3.9[a]	0.12[b]	1.8 ± 39.6[a]	1.6 ± 2.0[b]	75.5 ± 145.0[b]	-0.014 ± 0.22[a]	0.36 ± 1.46[a]

Table 2-4. Analysis of Covariance (ANCOVA) for differences between upstream and downstream values in 16 reaches of Migina catchment (Munyazi sub-catchment), Rwanda, of total suspended solids (TSS), total nitrogen (TN), total phosphorus (TP), temperature (Temp), pH, electrical conductivity (EC) and dissolved oxygen concentration (DO). Numbers are the mean squares for the response variables of the effects of land use/land cover (LULC, categorical variable), discharge (Q, continuous variable, ln-transformed) and their interaction. Significance of effects is indicated by *** (P<0.001), ** (P<0.01), * (P<0.05), ms (marginally significant, P<0.10), ns (not significant, P>0.10).

Explanatory variable	df	Response variable mean squares (MS)						
		Temp	PH	EC	DO	TSS	TP	TN
LULC	3	60.94^*	1.1454^*	22962^{***}	45.22^{***}	565800^{***}	0.02911^{ns}	13.23^{ms}
ln(Q)	1	0.19^{ns}	0.0310^{ns}	50482^{***}	9.74^{ns}	244702^{***}	0.01156^{ns}	100.63^{***}
ln(Q)*LULC	3	3.63^{ns}	0.8483^*	18703^{***}	1.27^{ns}	157215^{***}	0.03930^{ns}	43.70^{***}
Residuals(df)		17.30(161)	0.3147(166)	2530(164)	3.59(147)	19188(199)	0.06325(199)	6.03(199)

The ANCOVA (Table 2-4) showed significant interactions between the effects of LULC type and total monthly discharge, indicating that the relationship between discharge and water quality change was not the same for all LULC types. Upstream-downstream differences (higher downstream values) in TSS concentrations increased significantly with discharge in rice and vegetable reaches. Similarly, EC was increasingly higher downstream from vegetable reaches (linear regression, see Figure 2-10). The only other significant relationships between discharge and water quality change were for TN and EC in ponds/reservoir reaches (increasingly lower downstream).

2.4 Discussion

The objective of this study was to explain variation in water quality in relation to river discharge and land use and land cover (LULC). This was done by comparing water quality through different seasons, within and among three sub-catchments and through a more detailed study among river reaches with different LULC within one sub-catchment. Some additional data was gathered during two rainfall events. The data thus represent different spatial and time scales. Following on the hydrological studies of Munyaneza (2014), this is the first comprehensive study on water quality and land use in the Migina catchment. While this is a unique watershed with unique responses to environmental and anthropogenic pressures, the results are of importance for many similar catchments and valley bottoms in Rwanda and other parts of Africa that are currently under pressure of conversion for economic development.

Catchments with the same rainfall, but with different geology and land use can have different runoff. Akagera had a higher discharge than Munyazi and Mukura, as observed in 2009-2010 (296.9, 65.0 and 60.3 mm y^{-1} or 0.30, 0.11 and 0.08 m^3s^{-1}, respectively; Munyaneza 2014). Water abstraction for rice farming in Munyazi and Mukura and for drinking water supply in Mukura, and the higher degree of imperviousness in Akagera (Table 2-1) can also affect these patterns. Van den Berg and Bolt (2010) suggested that higher evaporation in Munyazi could cause lower discharge.

Catchment runoff influences water quality in different ways. The lower discharge in Mukura provides a plausible explanation of the higher water temperature there, while the systematic decrease in pH at higher flows (Figure 2-5) could be attributed to low pH buffering provided by ferralitic and histosols soils overlying geology of granitic, schistose and metamorphic rocks. High flows in the three sub-catchments were associated consistently with higher TSS concentrations (Figure 2-5e), and particularly evident during the two high flows of March and April 2013 (Figure 2-7f). That only some high flows (April, June and November 2011) were associated with higher TN or TP concentrations (Figure 2-5), reflects the complexity of TN and TP relationships with discharge. In the 16 more intensively studied reaches of the Munyazi sub-catchment (Figure 2-9), and for the two high flows (Figure 2-7), the non-linear relationship between TN and TP and discharge can be attributed to other factors including LULC (Hecky et al., 2003; Kingdon et al., 1999; USEPA, 2002).

The short-term (within one day) changes in temperature, pH, EC and DO during the rain events all showed decreases of these variables with increases in discharge (Figure 2-7). TP and TN concentrations increased rapidly with increasing discharge, that were maximal between 5 and 10 $m^3 s^{-1}$, then decreased until the peak discharge (Figure 2-7h, g). This corroborates the findings by Ijaz et al. (2007) and Bartley and Speirs (2010) that large events may carry the highest loads, but may not have the highest concentrations. Haygarth and Jarvis (1999) characterized the hydrology as "the most important factor providing the carrier and energy for nutrient transfer". The decrease in DO with increasing TSS (Figure 2-8) may be caused by the oxygen demand of sediments (Giga and Uchrin, 1990; Jason et. al, 2009).

The influence of LULC on water quality was clearly evident in the 16 studied reaches of the Munyazi catchment. High EC, TSS, TP and TN observed downstream of the urban area of Karubanda, and those dominated by rice and vegetable farming, were in marked contrast with low EC, TSS, TP and TN observed in the headwaters of Ibisi bya Huye and upstream land cover of grass/forest. The lower temperature and DO observed downstream of grass/forest-dominated reaches can be attributed to physical shading developed by the dense canopy of wetland grass/forest and by strong interaction with the deep groundwater whose temperature and DO are lower (van den Berg and Bolt, 2010). The significant differences between LULC types of water quality change (Table 2-3) suggest that rice and vegetable farming generate or facilitate the transport of higher TSS concentrations as opposed to ponds/reservoirs and grass/forest cover. For TP and TN, the picture is less clear.

The observed variation in water quality variables with discharge can be interpreted as a "build-up, washout and dilution mechanism". During base flow conditions, urban and agricultural areas, wetland grass and dams act as sinks for sediments and nutrients, which are released (flushed out) during the early stages of high flows (Bolstad and Swank, 1997; Hirsch, 2012). In the Migina catchment, nutrient build-up occurs at the start of the dry season (June-September) until the middle of the main agricultural season in October. Washout occurs in early periods of high flows and from the middle of the main agricultural season in November-December. Dilution occurs during the late stages of high flows and at the end of the high rainfall season (May and June). Different LULC types seem to modify this general pattern. Agricultural activities and waste disposal make more sediment and nutrient available for washout, enhancing the source function of a reach whereas standing water in ponds and reservoirs may increase the sink function of a reach. Higher TSS, TN and TP characteristic of the main farming season (September-January; Figure 2-9e, f, g) are consistent with the working of the soil by hoeing, levelling and weeding and associated irrigation and fertilizer application (Hecky et al., 2003; Jennings et al., 2003; Verhoeven et al., 2006; Ijaz et al., 2007). Dredging of fishponds can also transfer sediments to downstream waters. Jennings et al. (2003) reviewed a number of studies demonstrating increased phosphorus emmisions associated with high flows in agriculturally dominated catchments.

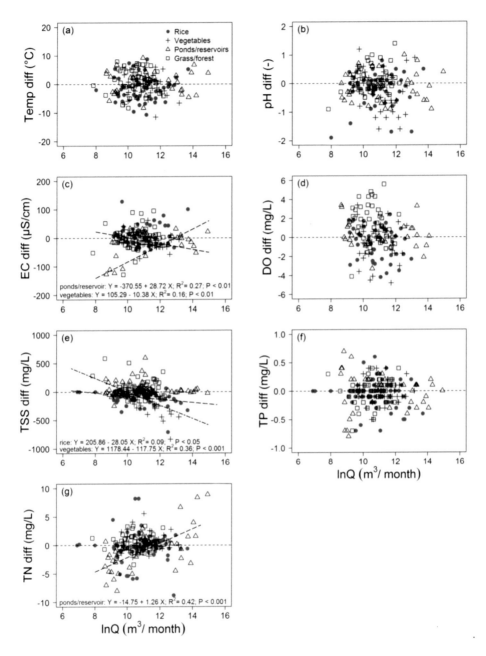

Figure 2-10. Differences between upstream and downstream water quality in relation to total monthly discharge (in m³/month, natural logarithm transformed) in 16 reaches with four land use/land cover (LULC) types in Migina catchment, southern Rwanda. For further explanation see text.

Separating seasonal variation in discharge and interrelated LULC is difficult. In the Rwandan catchments of this study land preparation and planting are done at the onset of the rainy season. Rainfall determines land use activities, and land use significantly influences hydrological processes like runoff, infiltration, evapo-transpiration and precipitation (Giertz et al., 2005; Scanlon et al., 2007). Significant interactions were found between the effects of LULC and discharge (Table 2-4; Figure 2-10), but because of large variance the regression analysis was only significant for TSS in rice and vegetable reaches, TN and EC in ponds/reservoir reaches, and EC in vegetable reaches. Nevertheless, these are clear indications that different LULC types influence the water quality changes in river reaches in different ways.

The large variations in TSS, TN and TP at the catchment scale and across all the study reaches and the non-linear relationships between discharge and nutrients (TN and TP) for the two high flows suggest that dynamics in land use change may not be captured by the analysis. While dominated by rice farming, the valley bottoms classified as "rice" are subject to different stages of land preparation, fertilizer application, crop growth, harvesting and fallow, and temporary conversion to other uses. Some human activities and land use can have major impacts on water quality over short periods. These include wastewater discharge, brick making, and grazing. Nutrients, especially nitrogen can also be strongly influenced by in-stream processes and washing, bathing, irrigation return flows, point and diffuse sources of pollution can all affect relationships between discharge and TP and TN. The monthly DO data are also insufficient to reliably interpret upstream-downstream changes in water quality in relation to LULC. Future field measurements should include more frequent simultaneous measurements of DO, pH, conductivity and temperature to characterize the impacts of different agricultural practices (e.g. by deployment of water quality sondes for 1-2 week periods). Other factors that may influence these relationships include the groundwater characteristics and the areas influencing water quality of the various reaches. Detailed investigation of area-specific retention of nutrients and sediment is discussed in a subsequent paper (Uwimana et al., 2018).

In the global picture of pollution, this study showed that farming activities and high flows can mobilize high concentrations of sediment, TN and TP, exceeding the guidelines set for many surface waters (e.g., RNRA, 2012). TSS far exceeded the maximum allowable concentrations (30 mg L^{-1}) most of the time and could reach 2601 mg L^{-1} during land preparation and or high flows. This was also observed for TN and TP concentrations. In some periods of base flows following higher flows, TN and TP were not detectable (< 0.001 mg L^{-1}), but exceeded the maximum allowable concentrations of some surface waters that are protected or considered sensitive (3 mg L^{-1} for TN and 0.1 mg L^{-1} for TP) during extended dry periods, higher flows, and farming periods.

In sub-Sahara Africa, agriculture dominates national economies and GDPs. With the increasing population and need for food security, pressure on land will force farmers to cultivate more areas of natural ecosystems like forests and wetlands, further degrading water systems (water quantity and quality), livelihoods and economies (FAO, 2011b). For the sustainable improvement of people's livelihoods, it is important to promote farm level adaptation strategies (Kotir, 2011) that balance livelihoods with ecosystem health. As rice and vegetable farming make valley bottoms more vulnerable to loss of sediments and possibly nutrients at high discharge, the combined effect of future climate change with conversion of more valley bottom wetlands to farming presents a further risk for water quality in river systems like the Migina. Extreme events, such as droughts and floods, are expected to increase in intensity and frequency in sub-Saharan Africa (FAO, 2011a) as a result of which stronger build-up and washout of nutrients can be expected. It is important that programmes to convert wetlands for agricultural exploitation take into account the importance of other LULC types, such as ponds and reservoirs (standing water) and grass/forest (natural vegetation) for catchment nutrient and sediment buffering. Safeguarding a certain percentage of a catchment as semi-natural or low-intensity land use can mitigate environmental impact, but requires not only effective national policies, but the local knowledge to incorporate this into operational land-use strategies (Verhoeven *et al.*, 2006). This, in turn, depends on clear communication, that alerts farmers and extension workers to the risks that a short-term gain in agricultural production can compromise longer term food security. In the face of increasing poverty and hunger, this is a major challenge, not only in Rwanda but across much wider areas of the continent.

2.5. Conclusion

In the three sub-catchments and across the 16 reaches of Migina catchment, water quality varied systematically with discharge and LULC. Concentrations of suspended solids were higher during high flow periods; while other water quality parameters (DO, pH, EC) were lower during high flows. TN and TP followed a mechanism of build-up in the dry season, washout in early periods of high flows and dilution at the late stages of rainy periods. The data suggest that agricultural activities makes reaches more vulnerable to the loss of sediments and possibly nutrients, whereas ponds or reservoirs and grass-forest cover enhance the sink function of reaches. With the increasing rates of population and need for food, pressure on land will continue forcing farmers to cultivate more areas of forests and wetlands, further degrading water systems (water quantity and quality), livelihoods and economies. To improve people's livelihoods and economic development in a sustainable way, farm level adaptation strategies need to be promoted that safeguard the ecosystem health on which livelihoods ultimately depend. More detailed analysis of the data, taking into account seasonal variation in land use and area-specific retention of reaches is needed to reduce variation and allow more firm conclusions.

2.6 Acknowledgements

This study was conducted as part of a cooperation between the University of Rwanda (former National University of Rwanda) and IHE Delft Institute for Water Education, The Netherlands. Financial support was provided by NUFFIC, The Netherlands through the NICHE programme and the Netherlands Fellowship Programme.

3. Effects of agricultural land use on sediment and nutrient retention in valley-bottom wetlands of Migina catchment, southern Rwanda[2]

Abstract

Factors affecting the retention and export of water, sediments (TSS), nitrogen (TN) and phosphorus (TP) were examined in the Migina river catchment, southern Rwanda from May 2012 to May 2013. Landscape characteristics and seasonal changes in land use and land cover (LULC), rainfall, discharge, and area-specific net stream yields of TSS, TP and TN were measured monthly in 16 reaches of the Munyazi sub-catchment with five valley bottom LULC categories (grass/forest, ponds/reservoirs, ploughed, rice, and vegetables). Valley bottoms dominated by grass/forest and ponds/reservoir types were generally associated with positive net yields of nutrients and sediments, while those with agricultural land covers had a net negative yields, resulting in net export. Water was retained only in reaches with ponds/reservoirs. Seasonally, there was a strong relationship between net yield and discharge, with 93%, 60% and 67% of the annual TSS, TP and TN yields, respectively, transported during 115 days with rain. During low flow periods, all LULC types had positive net yields of TSS, TP and TN (suggesting retention), but during high flow periods had negative net yields (suggesting export). Significant effects of hillside land use on sediment and nutrient yields were also found. Stream and river water quality in Rwandan valley bottoms are at risk of further deterioration due to declining natural ecosystems (grassland and forest) and increasing agricultural and urban development. It is important to adopt appropriate land management practices (minimal tillage, optimization of water use, strategic implementation of retention ponds and vegetation buffer zones) to intercept TSS, TP and TN in runoff from storm water and agricultural areas. Special attention is needed for critical periods of the year when farming activities (e.g. land preparation, fertilizer application) coincide with high flow events.

Keywords: water quality; catchment management; agriculture; land use; sediment retention; nutrient retention; wetlands

[2] Published as:
Uwimana, A., van Dam, A.A., Gettel, G.M., Irvine, K., 2018. Effects of agricultural land use on sediment and nutrient retention in valley-bottom wetlands of Migina catchment, southern Rwanda. Journal of Environmental Managmement 219, 103-114. https://doi.org/10.1016/ j.envman.2018.04.094.

3.1 Introduction

In Sub-Sahara Africa, land degradation threatens many livelihoods, and is estimated to cost 3% of Africa's annual agricultural gross domestic product (Jansky and Chandran, 2004). In Rwanda, land degradation and associated soil and nutrient losses reduce the capacity to feed the nation by 40,000 Rwandans each year (NISR, 2008). Soil loss was estimated at 10 t ha^{-1} y^{-1} or 120-136 kg of nitrogen, phosphorus and potassium (NPK) ha^{-1} y^{-1} (Henao and Baanante, 1999; World Bank, 2005). Erosion and nutrient loss also leads to water quality problems, and is the main source of nutrient and sediment pollution in surface water, groundwater and wetlands (UN-WWAP, 2009).

In East Africa, erosion and nutrients from fertilized farms and urban areas are the major causes of eutrophication and ecosystem degradation of Lake Victoria. The Migina river in the uppermost catchment of the Nile River is a tributary of the Akagera River, the main surface inflow to Lake Victoria. The typical catchment landscape consists of a series of rugged medium-high to high steep mountains, draining into narrow valley bottom wetlands. With an average population density about 450 persons km^{-2}, pressure on land in the Migina catchment is high, and soil degradation widespread. Conversion of valley-bottom wetlands for agricultural production, that has being ongoing for a number of years, is set to increase further as a consequence of the Rwandan government's food security policies (MINAGRI, 2009; 2010a). While effects of soil loss and nutrient exports on the eutrophication and sedimentation of rivers and lakes like Akagera River, Lake Cyohoha, as well as Lake Victoria are already apparent, there is a need for more detailed knowledge on how current land use practices affect water quality, and how valley-bottom wetlands can mitigate that.

The contribution of valley-bottom wetlands to water quality regulation depends on the complex interactions between hydrological flow paths from the hillsides to the streams (surface runoff and groundwater flows) and biogeochemical processes (Burt and Pinay, 2005; Lohse et al., 2009). Nutrient retention by floodplain wetlands through ecological and biogeochemical processes has been demonstrated in European and American wetlands (e.g. Johnson et al., 1997; Verhoeven et al., 2006). Little is known about the role of wetlands in African catchments (Dunne, 1979; Hecky et al., 2003; de Villiers and Thiart, 2007), where wet and dry seasons are more pronounced and where land-use conversion sometimes follows seasonal patterns which makes the generalized land use classifications that are commonly used less appropriate. A water quality study in Migina catchment showed that the relationship between water quality and discharge is influenced by the land use in the valley bottom, with reaches dominated by rice and vegetable land covers more prone to sediment and nutrient loss than those with intact wetland vegetation or water bodies like ponds or reservoirs (Uwimana et al., 2017; Chapter 2). However, that study only looked at water quality differences and did not consider the size and characteristics of reaches on sediment and nutrient retention.

The strong seasonal and spatial variation in the Migina valley-bottom farming systems is likely to have an impact on sediment and nutrient dynamics. The agricultural land passes through different land cover stages seasonally, as farmers follow a rotation of land preparation, planting, weeding, fertilizing and harvesting. Studies in other parts of the world have shown that the chemistry of streams is influenced by both hydrological pathways (from hillslopes through floodplains to streams) and the chemical and biological processes of soils and vegetation (Burt and Pinay, 2005). Sediment and phosphorus (P) transport generally depend on overland flows, while nitrogen (N) transport is more related to subsurface flows (Jordan *et al.*, 1997; Pärn *et al.*, 2012). In agricultural catchments, both N and P can be washed into surface waters by overland runoff shortly after the application of fertilizers and manures, or during livestock grazing. N transport through groundwater, often in the form of nitrate, is slower (Howden *et al.* 2011). Factors determining the movement of N, P and sediments include frequency and intensity of rainfall, land use and land cover, soil type (hydraulic conductivity and erodibility), slope length and angle, and processes like leaching, adsorption, and denitrification (Burt and Pinay, 2005; Lohse *et al.* 2009; Pärn *et al.*, 2012).

Field measurements of these processes are absent in Rwanda and generally scarce in Africa, where many catchments are un-gauged, sampling sites are difficult to access and water quality is not monitored regularly. In the Migina catchment both land use and stream flow are subject to strong seasonal fluctuations, and a better understanding of how these relate to land use patterns in driving sediment and nutrient dynamics is important for supporting catchment planning. This is relevant not only in Rwanda but in many other African countries with similar landscape features and similar challenges of food security, water quality degradation and climate variability. The overall objective of this study is to relate trends in the retention of total suspended solids (TSS), total phosphorus (TP) and total nitrogen (TN) to valley bottom land use & land cover, in the context of seasonal rainfall and landscape features in the Migina catchment, Rwanda. Specific objectives were to (1) describe landscape features (catchment area, soil, slope and population density) and quantify changes in seasonal valley bottom land use & land cover dynamics in selected reaches of Migina catchment; (2) assess seasonal variation in the rainfall and fluxes of TSS, TP and TN; and (3) examine the relationship between TSS, TP, and TN yield and land use & land cover in the valley bottoms.

3.2 Material and Methods

3.2.1 Study area and period

Migina catchment (Figure 3-1a,b) is located in Rwanda's Southern Province. It has an area of 261 km^2 and is shared between Huye, Gisagara and Nyaruguru Districts (29°42'- 29°48' E; 2°32'- 2°48' S). The geology in the area is typical granites, quartzites and schists. Weathering and ferralisation of these rocks has resulted in ferrallitic soil with clay deposits in the valley bottoms. The soil on the steep hillsides is composed of clayey and sandy material typical of ferralitic soil with good drainage, while the valley bottom soil consists of clay intermixed with organic detritus typical of Histosoils (Twagiramungu, 2006; Van den Berg and Bolt, 2010). Migina river is 40.4 km long and fairly straight in most of its course, with a channel of about 1 m deep and 1.5 m wide in the upstream to 3 m deep and 4 m wide downstream. A long-term average runoff coefficient of 0.25 was estimated for Migina catchment (Munyaneza *et al.* 2011), and during two rainfall events more than 80% of the total discharge was generated by subsurface flows (Munyaneza *et al.* 2012). Maximum altitude in the catchment is 2,247 m, with the valley floors ranging from 1,400 to 1,900 m. The climate in the catchment is temperate tropical humid, influenced by the proximity to the equator at 2° to the north, by Lake Victoria in the East, and by the altitude, with a pronounced dry period between June and September. Average temperature was 20 °C with low thermal amplitude. Rainfall distribution in 2009-2011 was fairly uniform across the catchment (coefficient of variation of 7%), with annual rainfall about 1,200 mm (Munyaneza *et al.*, 2014).

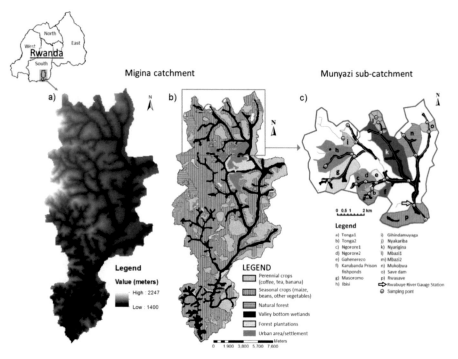

Figure 3-1: a) Digital Elevation Model (source: Centre for Geographic Information Systems & Remote Sensing, University of Rwanda, 2008); b) Migina catchment; and c) 16 reaches of the upper Migina river in the Munyazi sub-catchment, southern Rwanda (b). Table 2 shows the land use & land cover (LULC) characteristics of these reaches.

Agriculture dominates land use and land cover (LULC) in both the valley bottoms and hill slopes. Valley bottom LULC includes rice farms, vegetables (maize, beans, potatoes, carrots, tomatoes), clay, silt and sand mines, grassland, and fishponds and dams (standing water). Between crops, plots may be left fallow or ploughed. On the hill slopes, bananas are the main crop and cover the soil more or less the whole year. Land cover within the crop growing season in the valley bottom can change from completely bare soil resulting from hoeing to a dense canopy at the end of the season. Sixteen river reaches with different LULC combinations in the valley bottoms were selected in the Munyazi sub-catchment (Figure 3-1c) because it represents all major LULC types in Migina catchment. Seasonal LULC patterns in the valley bottom land surrounding each reach were monitored monthly in the period May 2012 - May 2013 along with the TSS, TP and TN concentrations before entering and after passing through the valley-bottom lands (Table 3-1). In the following text, the term "reach" is used for a river section with the associated valley-bottom land.

Table 3-1. Landscape characteristics of the study reaches in Munyazi sub-catchment, Migina river, southern Rwanda (see also Figure 3-1).

Reach	Valley bottom				Hillside		Pop. density (km^{-2})
	Length (km)	Slope (%)	Area (ha)	Reach/ valley area ratio (%)	Slope (%)	Area (ha)	
Ngorore-2	0.80	0.5	8.94	9.8	15.2	82.10	155
Tonga-1	0.48	1.4	3.97	7.9	15.7	46.46	64
Gihinda	1.04	2.2	5.00	9.5	10.7	57.89	510
Nyarigina	4.94	0.9	104.6	17.0	11.2	509.48	2065
Mukobwa	1.74	1.2	25.62	8.7	10.3	267.65	629
Gahenerezo	1.42	0.4	10.93	15.7	11.7	58.67	2326
Masoromo	2.96	1.4	51.52	13.4	15.7	331.90	484
Ibisi	2.11	2.5	10.46	2.7	18.0	372.26	354
Nyakariba	1.29	0.8	15.43	16.9	14.5	75.76	670
Ngorore-1	0.72	1.5	11.58	13.2	14.4	76.11	72
Mbazi-1	0.48	1.5	3.49	8.2	10.6	39.22	26
Mbazi-2	0.45	1.4	7.50	14.2	11.0	45.47	53
Tonga-2	0.36	1.4	5.71	14.0	15.7	35.07	64
Rwasave	1.53	0.1	45.56	11.2	11.8	360.64	761
Save dam	0.69	0.8	6.31	8.0	11.0	72.50	1992
Karubanda	0.37	1.1	7.29	12.7	11.7	40.00	4480

3.2.2 Data collection

3.2.2.1 Reach characteristics, population density and land use

A Digital Elevation Model (DEM, 10x10 m resolution, Figure 3-1a) produced by the UR-CGIS (Centre for Geographic Information Systems & Remote Sensing, University of Rwanda) in 2008 was used to delineate the study area and identify the hydrological networks, determine the reach areas, slopes and length of the valleys. Hydro-geomorphic modifications in the sub-catchment exist in the form of reservoirs (Save Dam), fishponds (Rwasave fish farm with 103 fishponds; and Karubanda fish farm with 10 ponds) and small water intake points. Soil types were classified using field observation and literature review (Verdoodt and van Ranst, 2006). The population density in each reach was determined by interpolation based on demographic data from Huye district (NISR, 2014) and reach surface areas, assuming a uniform distribution of the population within cells. Every month, LULC in each reach was determined through a combination of visual observation, photography and overlaying of this information with Google satellites images (Google earth V6.2.2.6613, Butare, Rwanda. DigitalGlobe 2012. http://www.earth.google.com [May 20, 2012]) and the DEM. Each month, the valley bottom area covered by rice farming, vegetable farming, grassland/forest, fishponds/reservoir, or ploughed land was measured in each reach. Based on these observations, five LULC categories were determined, and the proportion of each were estimated monthly in each reach. Hillside LULC in each reach was classified each month into four categories: ploughed land, vegetables, urban area, or forest/grasses, according to the predominance of each LULC type in each reach. The urban area, settlements, commercial and institutional developments were located on the top of hills, while forest/grass were located on the hillside. Some reaches had unique hillside LULC characteristics, notably Huye aerodrome in Tonga-2, the University of Rwanda in Rwasave, the Catholic University and many boarding schools in Save, Karubanda prison in Karubanda FP, a densely populated centre in Gahenerezo, and undisturbed forest upstream Ibisi. The slope, soil type, reach length, reach area, and population density were assumed to be constant during the study period.

3.2.2.2 Rainfall, discharge and water quality measurements

Discharge, TSS, TP and TN were measured monthly at sampling points upstream and downstream of each reach (Figure 3-1c). Rainfall was recorded using a rain gauge installed at the Rwasave Fishpond Station (long. 2.602°, lat. 29.757°, alt. 1665 m). Discharge was measured monthly at each site using the area-velocity method with 20 cm sections across the stream. Water depth in each section was measured using a staff gauge, and water velocity using a propeller flow meter (Universal Current Meter, OTT C31, Hydromet GmbH, Kempten, Germany). Stream discharge was calculated as the sum of discharges (product of cross section area and water velocity) in each section. As the monthly measurements were not able to capture the short term daily flow variations, the measured discharges were related to the corresponding water pressure data from the Rwabuye river gauging station (Figure 3-1c), which recorded every 30 minutes using an automatic water level logger (Mini-Diver DI501, Schlumberger Water Services, Delft, The Netherlands). The resulting regression equations (exponential or linear, with R^2 of 0.84-0.96) were used to estimate 30-minute interval discharges for each reach. Monthly discharge (m^3 month^{-1}) at each site was calculated by summing up these estimated 30-minute discharges.

Water samples were collected at different depths (surface, middle and bottom) using Ruttner Van Dorn bottles (Hydro-Bios) and mixed to form a composite sample of the site. TSS was determined on-site using a portable colorimeter DR/890 (HACH, Colorado, USA). Samples for TP and TN were preserved using sulphuric acid and transported to the laboratory of the University of Rwanda (Huye campus) to determine TN and TP using the persulfate digestion method (APHA, 2005). Other water quality parameters, including specific electrical conductivity (EC), dissolved oxygen concentration (DO), pH and temperature were also collected but results are reported elsewhere (Uwimana et al., 2017; see Chapter 2).

To distinguish days with event flow from days with base flow, the strong relationship between the EC of the water and flow at the Rwabuye gauging station measured during two flow events was used (regression with $R^2 = 0.75 - 0.80$; Uwimana et al., 2017). Days on which discharge had EC values < 90 μS cm^{-1} were classified as event flows, assuming that stream water was dominated by the direct rainfall runoff. Based on the Rwabuye records, there were 115 days of event flow and 250 days of base flow during the experimental period. Using these values, the proportion of annual yields transported during event flow or base flow were calculated.

3.2.2.3 Yield calculations

Because TSS, TP and TN were measured on a monthly basis, daily concentration (g m^{-3}) estimates were made from regressions between discharge (independent variable) and TSS, TP and TN (dependent variables) using a method described in Wollheim et al. (2005). First, regressions were estimated for each reach between monthly TSS, TP and TN concentrations and the corresponding discharge at the Rwabuye gauging station. These regressions were then used to predict daily TSS, TP and TN values at each site from the daily discharge measurements at Rwabuye. This method was effective during event flow ($R^2 > 0.80$) for most of the sites, but was not effective at base flow, which gave low R^2 values (<0.2). Therefore, dry season fluxes were calculated based on the measured TSS, TP and TN concentration in the same month.

Reach yields were calculated by multiplying concentrations of TSS, TP or TN with discharge (resulting in the flux) and dividing by the corresponding catchment area (resulting in the yield). The difference between upstream and downstream yield (from here referred to as 'net yield') was expressed in kg ha^{-1} month^{-1} and could take positive (retention) or negative (export) values. For water, this net yield was calculated as the difference between upstream and downstream monthly discharge (Q_u and Q_d at reach inlet and outlet, respectively) divided by the reach area (A) and expressed in mm month^{-1}, with positive values indicating water retention and negative values water export (or water yield).

3.2.2.4 Sediment and nutrient yields, land use and catchment characteristics

Monthly observations for the 16 reaches in the period May 2012 to May 2013 (13 months) resulted in a dataset with 208 observations. To analyse the net yield data, three approaches were explored. The first approach ("ANOVA approach") compared the net yields of water, TSS, TN and TP among the five LULC types using repeated-measures ANOVA with LULC-type, month and their interaction as fixed variables and an autocorrelation structure for month (Mangiafico, 2016).This approach aimed at an overall comparison of the five LULC-types over the whole study period.

The second approach (the "ANCOVA approach") compared the five valley bottom land use categories in terms of net yield of water, TSS, TP and TN, with discharge as a covariate and taking into account random variation among reaches (Zuur et al., 2007; 2009). This approach allowed for a more detailed analysis including variation caused by differences in discharge and by variability among individual reaches. Discharge was log-transformed (base 10) to stabilize the variance. First, a conventional linear model with no random reach effect was estimated (Model 1; this is equivalent to a traditional ANCOVA model). Then, reach was added both as a random intercept (Model 2) and random slope (Model 3). Finally, models allowing different residual variances for each reach (Model 4) and a combination of Models 3 and 4 (Model 5) was estimated. The resulting models were compared using the Akaike Information Criterion (AIC) and residuals were plotted to check for homogeneity.

The third approach ("regression approach") related differences in monthly net yield of TSS, TP and TN to the discharge, reach catchment area (ha), valley bottom slope (%), hill slope (%), population density (individuals/ha), and the proportion (%) of LULC observed each month for each of the LULC categories. This analysis tried to identify reach characteristics that were responsible for differences in net yield of sediment and nutrients. For this, linear models using discharge (log-transformed) and all other possible explanatory variables were estimated (Zuur et al., 2007; 2009). First, the distribution and correlation of all variables were checked (Pearson correlation) to avoid dependence and collinearity. Then, a conventional linear model with no random reach effect was estimated (Model 6). Subsequently, models were estimated with a variance structure to allow each reach a different variance (Model 7), with a random intercept and slope for reach (Model 8), or a combination of these (Model 9). Finally, models allowing different variance per reach that increased with discharge were estimated (Model 10). Significance of effects and residual distribution were checked iteratively. The "best" model was selected based on a comparison of the AIC and on residual plots.

Spatial analyses were done using ArcGIS version 9.3.1 (ESRI Inc., Redland, CA, USA). Models were fitted using R version 3.3.1 (R Core Team, 2016), using the functions lm (stats package), gls and lme (nlme package; Pinheiro et al., 2017). The autocorrelation for month was defined using corAR1 within the gls function. Values for autocorrelation were estimated using the ACF function in library nlme. Variance structures for reaches were defined using varIdent, and to calculate different variances per reach that increased with discharge varComb was used, both in package nlme.

3.3 Results

3.3.1 Reach characteristics, population density, land use and reach classification

The catchments were dominated by hillsides, with the valley bottoms occupying only 2.7% (in Ibisi) to 17% (in Nyakariba) of the catchment area (Table 3-1). The length of the valleys ranged from 0.36 km (Ngorore-1) to 4.94 km (Nyarigina). The slopes were gentle in the valleys (0.13-2.45 %) and steep on the hillsides (10-18%), and decreased from upstream to downstream in both valley and hillside. The population density ranged from 26 km^{-2} in the rural area (Mbazi-1) to 4,500 km^{-2} in the urban area (Karubanda), which houses a prison and schools (Table 3-1).

Figure 3-2 shows seasonal LULC dynamics in the sub-catchment. In valley-bottom wetlands (Figure 3-2a), rice farming was the dominant activity occupying 26-47 % in two growing seasons (February-July and August-January). Vegetable farming occupied 18-28 % in three growing seasons (September-January, February-May and June-August). Ploughed land occupied 12-32 % of the valley bottom and was generally inversely proportional to the land occupied by rice and vegetables, indicating periods of land preparation or harvest, or sometimes clay harvesting. Ponds/reservoir and grass/forest did not vary within seasons, occupying the lowest proportion of LULC of about 6% and 10%, respectively.

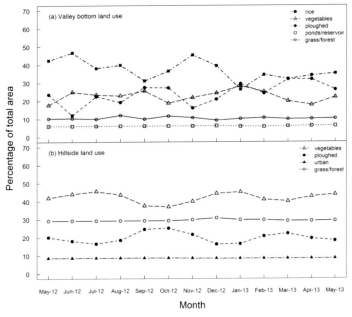

Figure 3-2: a) Seasonal dynamics of land use and land cover (LULC) in valley-bottom wetlands; and b) LULC on associated hillsides , expressed as percentage of total land use/cover in the entire study area (16 river reaches in Munyazi sub-catchment, Migina, southern Rwanda). LULC categories were vegetables, rice, ploughed, grass/forest, ponds/reservoir and urban.

Vegetable farming dominated 37-45% of hillside LULC (Figure 3-2b). Ploughed land occupied 16-25% and like the valley bottoms, was inversely proportional to vegetable cover. A higher proportion of ploughed land was observed between September and October and between February and April during hoeing and vegetable planting. Grass/forest occupied 29.5 % of the hillsides and did not vary with season. The urban area was also constant during the study period and, at 8.5% cover, was the least abundant LULC on hillsides.

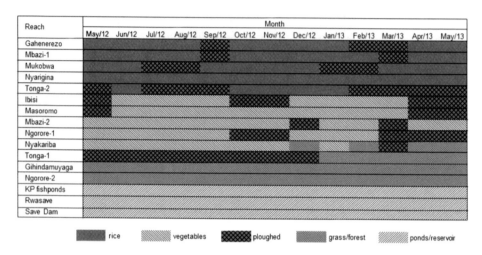

Figure 3-3. Dynamic classification of river reaches into five land use & land cover (LULC) categories based on dominant LULC in every month of the period May 2012 - May 2013. LULC categories were rice, vegetables, ploughed, grass/forest, and ponds/reservoir. For more explanation, see text.

Figure 3-3 shows the 'dynamic' classification of reaches in five LULC categories, taking into account the seasonal LULC changes among them. Three reaches had permanent fishponds (Rwasave, KP fishponds) or a reservoir (Save dam) and were classified as 'ponds/reservoir' throughout the study period. Similarly, Ngorore-2 and Gihindamuyaga were dominated by natural grass or forest vegetation and were classified as 'grass/forest'. The remaining reaches were dominated by agricultural land use and were classified as 'rice'or 'vegetables' for periods of 3-4 months, alternating with shorter periods of 'ploughed' or 'grass/forest' (Figure 3-3). In the Tonga reaches, clay mining resulted in large areas of bare soil, which were classified as 'ploughed'.

3.3.2 Rainfall, water , TSS, TP and TN yields at Rwabuye gauging station

Rainfall at Rwasave (close to Rwabuye gauging station) varied from 0 mm month^{-1} in July 2012 to 360 mm month^{-1} in April 2013 (Figure 3-4a). The seasonal rainfall pattern was similar to the pattern in water yield, which in turn showed similarity to the pattern of TSS, TP and TN yields (Figure 3-4b-d). For example, a TSS peak yield of 192 kg ha^{-1} d^{-1} at the Rwabuye river gauging station was observed on 5th April 2013 during the heaviest rainfall of

75 mm. The same event was associated with high TP and TN yields (0.02 kg TP and 1.68 kg TN ha^{-1} day^{-1}, respectively). On an annual basis, 93% of the annual TSS yield was transported during the 115 rainy days of the year. The other 7% were transported during base flows in the remaining 250 days. Compared with TSS yields, the proportions of the annual TP (60%) and TN (67%) yields transported during flow events were lower.

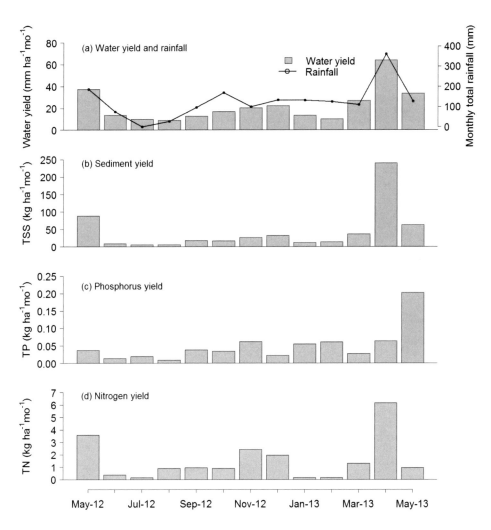

Figure 3-4. Seasonal dynamics of water yield (a: WY), load of total suspended solids (b: TSS), total phosphorus (c: TP) and total nitrogen (d: TN) at the Rwabuye river gauging station near the outlet of the Munyazi sub-catchment, Migina river, southern Rwanda. Data on rainfall (a) were collected at the Rwasave station.

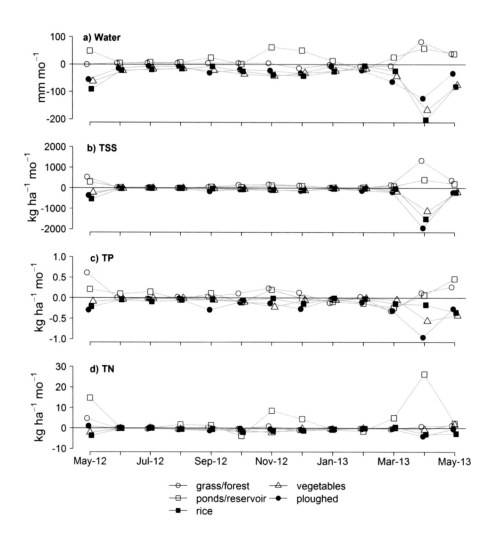

Figure 3-5. Net yield of water (a: Water), total suspended solids (b: TSS), total phosphorus (c: TP) and total nitrogen (d: TN) in 16 reaches of the Migina river (Munyazi sub-catchment), southern Rwanda. Data are means of reaches (n = 3-5) with different land use/cover, estimated monthly between May 2012 and May 2013. Results of statistical analysis are presented in Table 3.

Table 3-2: Predominance of land use & land cover (LULC) in valley bottoms and hillslopes of 16 reaches of the upper Migina catchment. Numbers presented are average percentage cover (coefficient of variation in parenthesis) for 13 monthly observations. Reaches are grouped according to dominant LULC in the valley bottom. See Figure 3-3 for detailed monthly LULC categorization.

Reach	Valley bottom LULC (%)					Dominant valley bottom LULC	Hillside LULC (%)			
	Ploughed	Rice	Vegetable	Water	Grass		Ploughed	Vegetable	Urban	Forest
Nyarigina	21 (63)	71 (19)	7 (58)	0	0	rice/ ploughed	23 (17)	47 (8)	10 (0)	20 (0)
Mukobwa	28 (113)	64 (50)	8 (28)	0	0		19 (46)	53 (13)	0	27 (16)
Mbazi-1	19 (91)	72 (24)	9 (41)	0	0		32 (47)	48(32)	0	20 (0)
Gahenerezo	36 (85)	55 (54)	8 (61)	0	0		15 (9)	15 (9)	25 (0)	45 (0)
Tonga-2	52 (66)	48 (72)	0	0	0		7 (55)	10 (33)	15 (0)	68 (5)
Mbazi-2	24 (122)	0	56 (49)	0	20 (91)	vegetables/ ploughed	18 (45)	52 (15)	0	30 (0)
Masoromo	37 (52)	0	63 (31)	0	0		23 (27)	57 (11)	0	20 (0)
Ngorore-1	45 (36)	0	55 (29)	0	0		23 (27)	27 (24)	30 (0)	20 (0)
Ibisi bya Huye	32 (34)	8 (40)	43 (20)	0	15 (0)		24 (36)	56 (15)	0	20 (0)
Save Dam	5 (158)	0	5 (80)	50 (16)	40 (0)	ponds/ reservoir	13 (35)	27 (16)	20 (0)	40 (0)
KP Fishponds	20 (76)	0	30 (49)	50 (0)	0		10 (0)	40 (0)	50 (0)	0
Rwasave	11 (60)	38 (21)	6 (61)	30 (0)	15 (0)		10 (0)	15 (0)	15 (0)	60 (0)
Gihindamuyaga	6 (27)	0	4 (60)	0	91 (3)	grass/ forest	25 (21)	40 (13)	5 (0)	30 (0)
Tonga-1	43 (49)	0	0	0	57 (37)		8 (50)	10 (54)	47 (5)	35 (0
Ngorore-2	6 (48)	0	26 (33)	0	68 (13)		26 (26)	39 (17)	0	35 (0)
Nyakariba	17 (77)	0	43 (30)	0	40 (0)		25 (32)	45 (17)	5 (0)	25 (0)

3.3.3 Net yield of water, TSS, TP and TN in relation to land use categories

Consistent positive values of net yield were observed in reaches dominated by grass/forest and ponds/reservoir, while negative values (export) dominated in agricultural reaches (ploughed, rice, and vegetables) (Table 3-3, Figure 3-5). Mean net yield of water over the observed period ranged from a low -60.5 mm month^{-1} in rice to -1.87 mm month^{-1} in grass/forest. Highest net yield of 28.1 mm month^{-1} was found in ponds/reservoir. Mean net yield of TSS was highest in grass/forest and ponds/reservoir (224.8 and 117.0 kg ha^{-1} month^{-1}, respectively) and lowest in ploughed reaches (-269.1 kg ha^{-1} month^{-1}). Mean net TP yield showed the same trend with high values in grass/forest and ponds/reservoir (0.050 and 0.094 kg ha^{-1} month^{-1}, respectively) and the lowest value in ploughed (-0.240 kg ha^{-1} month^{-1}). For TN, net yield was highest (4.64 kg ha^{-1} month^{-1}) in ponds/reservoir and lowest (-0.516 and -0.564 kg ha^{-1} month^{-1}) in reaches with vegetables and ploughed land, respectively. Generally, only the minimum and maximum values among the land use types were significantly different (p<0.05, repeated measures ANOVA; Table 3-3), while there were no significant differences among the months.

Table 3-3. Least square means of the net yield (upstream minus downstream) of water, total suspended solids (TSS), total phosphorus (TP) and total nitrogen (TN) for different land use categories in 16 reaches of the Munyazi sub-catchment in the Migina River, southern Rwanda. Numbers are means of 13 monthly net yield estimates. Means within a column with different superscript letters were significantly different (P<0.05; repeated measures ANOVA). Differences among months were not significant.

Land use category	Net yield			
	Water (mm month^{-1})	TSS (kg ha^{-1} month^{-1})	TP (kg ha^{-1} month^{-1})	TN (kg ha^{-1} month^{-1})
grass/forest	-1.87 [ab]	224.8 [b]	0.0503 [b]	0.408 [ab]
ponds/reservoir	28.1 [b]	117.0 [ab]	0.0940 [b]	4.638 [b]
Rice	-60.5 [a]	-246.9 [a]	-0.108 [ab]	-1.368 [a]
Vegetables	-27.6 [ab]	-91.6 [ab]	-0.0861 [ab]	-0.516 [a]
Ploughed	-34.5 [ab]	-269.1 [a]	-0.240 [a]	-0.564 [a]

3.3.4 Water, sediment and nutrient yield in relation to land use and discharge

Using the "ANCOVA-approach", the model with the lowest AIC was Model 5 for all four net yield indicators, except for TN where Model 4 had the lowest AIC (Table 3-4, Figure 3-6). The most informative outcome is the slope of the discharge-net yield relationship, as different slopes indicate different relationships between discharge and net yield among the LULC categories. Significant effects of LULC category (t-test, p < 0.05) were found for all net yield types. Ponds/reservoir showed the highest, and all positive, slope values for all net yield types except for TP where grass/forest was the highest (Table 3-4; Figure 3-6). In contrast, the reaches with vegetables and ploughed land had the lowest, and mostly negative slope values, indicating that at higher discharge values streams in reaches with these land uses were increasingly exporting water, sediment and nutrients. Rice and grass/forest had intermediate slope values. Rice had an intermediate position compared with all other LULC types with slope values closer to zero (i.e., neither negative nor positive) than the other LULC types.

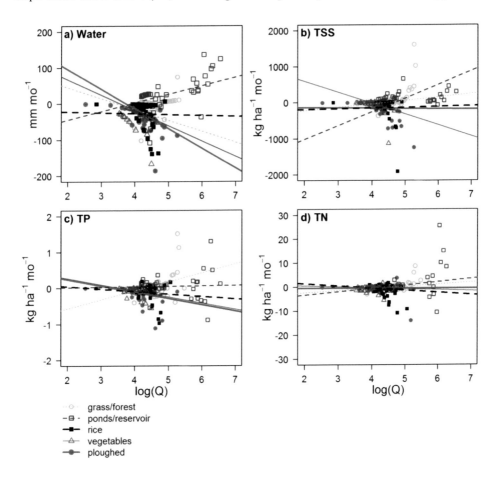

Figure 3-6. Relationships between discharge Q (in m^3 $month^{-1}$, ^{10}log-transformed) and net yield of water (a: Water), sediment (b: TSS), total phosphorus (c: TP) and total nitrogen (d: TN) in five different land uses in 16 reaches of the Munyazi sub-catchment, Migina river, southern Rwanda. Regression lines show the fixed effects of the mixed models presented in Table 4. Only lines with significant slopes (P < 0.05) are shown.

Table 3- 4. Results of mixed models[1] using the "ANCOVA" approach. Detailed results of the models with the lowest AIC (the "best" model, AIC-value in italics) are presented, including fixed effects (intercept and slope of the relationship between [10]log-transformed discharge and retention), and random variation (intercept, slope and residual). Significance of intercepts and slopes are indicated with ms (p<0.1), * (p<0.05), ** (p<0.01) and *** (p<0.001). Total no. of observations (N) was 208.

	Dependent variable (Net yield)						
	Water (mm month^{-1})		TSS (kg ha^{-1} month^{-1})		TP (kg ha^{-1} month^{-1})		TN (kg ha^{-1} month^{-1})
Model comparison[1]							
AIC (Model 1)	2244.25		3141.86		173.17		1304.27
AIC (Model 2)	2220.48		3143.75		173.69		1294.21
AIC (Model 3)	2128.51		3115.10		154.67		1242.55
AIC (Model 4)	1855.89		2721.69		-69.84		*756.74*
AIC (Model 5)	*1727.72*		*2691.24*		-78.73		759.20
Model 4/5 results							
Fixed effects:							
- intercept							
GrassForest	105.00	*	-408.44		-1.058	***	-3.942 Ms
PondsReservoir	-91.09	**	-1772.49	**	-0.002		-6.026
Rice	-16.49	**	-230.82		0.182	***	3.228 Ms
Vegetables	151.60		1205.89	**	0.620	***	1.355 *
Ploughed	204.69	*	-116.82		0.601	***	-0.405
- slope							
GrassForest	-30.14	**	93.95	Ms	0.248	***	0.858 Ms
PondsReservoir	23.56	***	378.53	**	0.012		1.374
Rice	-2.68	**	20.16		-0.068	***	-0.905 Ms
Vegetables	-42.41		-303.55	**	-0.172	***	-0.400 *
Ploughed	-54.61	*	-8.26	Ms	-0.176	***	-0.004
Random effects (stdev):							
Intercept	22.37		165.14		0.352		
slope log(Q)	0.17		0.09		0.099		
Residual	9.51		163.15		0.063		

[1]Model 1: conventional linear model with net yield as dependent variable and discharge ([10]log-transformed) and LULC category as independent variables, and no random effects (traditional ANCOVA model); Model 2: reach added as random intercept; Model 3: reach added as random intercept and slope; Model 4: as Model 1 but allowing different residual variances for each reach; Model 5: combination of Models 3 and 4.

Table 3-5. Results of linear mixed effects models, using the regression approach[1]. Detailed results of the models with the lowest AIC (the "best" model, AIC-value in italics) are presented, including fixed effects (intercept and slopes of the models with net yield of water, TSS, TP or TN as dependent variable), and random variation (intercept, slope and residual). Significance of intercepts and slopes are indicated with ms (p<0.1), * (p<0.05), ** (p<0.01) and *** (p<0.001). "Expon" is the exponent of the varComb variance structure in which different variances per reach are allowed, as well as an exponential increase of variance with discharge. Total no. of observations (N) was 208.

	Dependent variable (Net yield)						
	Water (mm month^{-1})		TSS (kg ha^{-1} month^{-1})		TP (kg ha^{-1} month^{-1})		TN (kg ha^{-1} month^{-1})	
Model comparison[1]							
AIC (Model 6)	2346.75		3237.00		197.81		1367.07	
AIC (Model 7)	1871.80		2741.69		-52.45		768.381	
AIC (Model 8)	2166.85		3237.81		173.73		1258.31	
AIC (Model 9)	*1715.62*		2743.69		-55.48		768.24	
AIC (Model 10)	1839.54		*2379.93*		*-172.25*		*654.82*	
Model 9/10 results							
Fixed effects:							
Intercept	-378.05	***	20.58266		$19.6 \cdot 10^{-3}$		0.282	***
LogQ	-28.24	***	-10.1409	***	$-12.2 \cdot 10^{-3}$	***	-0.201	***
Wforgrass	47.82	*	51.79	***	$7.50 \cdot 10^{-3}$		0.248	*
Hforgrass	-16.08		76.16	***	$54.2 \cdot 10^{-3}$	***	0.068	
Hslop	37.09	***	-2.21		$-1.52 \cdot 10^{-3}$		0.016	
Pop	-2.71	***	1.87	***	$2.13 \cdot 10^{-3}$	*	0.007	
Random effects:							
intercept (stdev)	0.01948		19.14		$0.127 \cdot 10^{-6}$		$2.066 \cdot 10^{-6}$	
slope Log(Q)	31.89		$17.3 \cdot 10^{-6}$		-		-	
Residual	5.44		$0.134 \cdot 10^{-6}$		$0.487 \cdot 10^{-6}$		$8.002 \cdot 10^{-6}$	
Expon	-		4.84		2.85		2.56	

[1]Model 6: conventional linear model with net yield as dependent variable and discharge ([10]log-transformed), proportion of natural vegetation in the valley bottom (Wforgrass), proportion of natural vegetation on the hillside (Hforgrass), hillslope (Hslop, in %) and reach population size (Pop) as independent variables, and no random effects; Model 7: variance structure added to allow each reach a different variance; Model 8: random intercept and slope effect added for reach; Model 9: combination of Models 7 and 8; Model 10: allowing different variances per reach that increased with discharge (Model 10).

With respect to the random effects of reaches the models showed mixed results. All random intercept and slope models were better than models without these random effects, indicating a significant contribution of individual reaches to variation in net yield. Some reaches were responsible for a large part of the overall variance (Figure 3-7a). For water, TSS and TP these were Tonga2, Gahenerezo and Gihindamuyaga, while for TN Rwasave caused a major part of the residual variance.

3.3.5 Water, sediment and nutrient yield and reach characteristics

After trying different combinations of explanatory variables that described the reaches, the following set of variables emerged for the fixed effects of the 'regression approach' models: discharge Q (log-transformed), the proportion of forest/grass in the valley bottom, the proportion of forest/grass on the hillside, the slope of the hillside, and the population density in the reach area (Table 5). For water, Model 9 gave the lowest AIC while for TSS, TP and TN Model 10 was selected. In all models, discharge had a significant ($p < 0.001$) negative effect, while the proportion of forest and grass in the valley bottom had a positive effect on net yields of water and sediment/nutrients (all significant at least $p < 0.05$, except for TP). The proportion of forest and grass on the hillsides also had a significant positive effect ($p < 0.001$) on net yields of TSS and TP but was not significant for water and TN. Hillslope surprisingly had a positive effect on water ($p < 0.001$) but was not significant for the other net yield indicators. Population size had a negative effect on net water yield ($p < 0.001$) and positive effects on TSS and TP yields ($p < 0.001$ and $p < 0.05$, respectively).

With respect to the random effects of reaches, there was a large slope effect for net yield of water. Variation increased with discharge for TSS, TP and TN yields. Most of the random effects were due to differences between individual reaches (Figure 3-7b). For water, this was mostly due to the reaches Gahenerezo and Gihindamuyaga, while for TSS, TP and TN the reach Nyakariba and, to a lesser extent, Gahenerezo and Mbazi2 were responsible for the variation.

3.4 Discussion

Clear differences in net yield of sediment and nutrients were observed among the different LULC categories. Considering the entire study period, stream reaches with agricultural activities on average exported sediments and nutrients while reaches with more natural vegetation retained them (Table 3-3). There were, however, seasonal differences that were related to rainfall, discharge, and land use. In months with higher flows, the export of agricultural reaches became higher, while in the more natural reaches (grass/forest) the effect of higher discharge was mixed, with more negative yields of nitrogen but positive yields of sediment and phosphorus. Reaches with ponds/reservoir showed higher water and nutrient yields and almost the highest net sediment yield, all showing a positive relationship with discharge (Figure 3-5). These differences among land uses in the valley bottom were underlined by the ANCOVA models, with positive slopes for TSS and TP in grass/forest and ponds/reservoir but negative slopes in the reaches with vegetables or ploughed land (Table 3-4, Figure 3-6). The observed seasonality was confirmed by the large proportions of TSS, TP and TN transported during rainfall events (93%, 60% and 67%, respectively). In all LULC types, TP and TN were mostly exported in the periods August-October and January-March (Figure 3-5, 3-6), both corresponding with the start of the main agricultural seasons just before periods of peak rainfall (Figures 3-2 and 3-4a).

Agricultural activities in the valley bottom only partially influence the overall transport of nutrients from hills to stream. Due to the high hydraulic conductivity of the ferralitic soils in Migina catchment, rainwater infiltrates rapidly to the groundwater and river flow is dominated by subsurface flow (Munyaneza et al., 2011; 2012). In the valley bottoms, the shallow groundwater layer (0.2-2 m) is separated from deeper groundwater by an impermeable red-brown subsoil layer (van den Berg and Bolt, 2010). Overland runoff from the hills is only generated during peak rainfall events. This runoff will be mostly responsible for the transport of sediment and associated P, while dissolved N is more likely to be transported by subsurface flows (Burt and Pinay, 2005). In the Migina catchment, this was confirmed by high nitrate concentrations found in some wells, attributed to human activities on the hillsides (van den Berg and Bolt, 2010).

Hydrological connectivity and the transport of water, sediment and nutrients is also influenced by the hillslopes (e.g. soil type, slope, land use and cover, waste discharge). Water yields in all valley bottom LULC categories were negative and decreased with increasing discharge, except for ponds/reservoir reaches which stored water as shown by a positive net water yield (Tables 3-3, 3-4). With the regression models, we tried to link net yields to some of the reach and hillside characteristics. The options for including variables in a model are limited because of co-variation (e.g., increasing agricultural land use automatically implies decreasing natural vegetation) and lack of detailed information about some variables. Other reach differences are not explicit in the analysis but can be understood by observing the characteristics of individual reaches. The model results seem logical in terms of valley bottom and hillside land use, with positive effects of less farming and more natural land cover on sediment and P net yields (Table 3-5). The effect of hillslope was not significant except for water, but the positive effect on net water yield is difficult to explain. Perhaps factors that were not included in the model played a role, such as terraces and trenches that are common in densely populated areas. The effect of population size on sediment and P net yields were positive, perhaps as a result of more paved surfaces and less erosion from urban centres, but the likely higher P discharge from wastewater could contradict such an explanation.

In the valley bottom, the fate of sediment and nutrients is determined by a combination of natural processes and human activities (including farming). At low flow velocity particulates, along with adsorbed phosphorus may settle. At higher flows, sediments are mobilized and move downstream. Natural depressions and river banks can create inundated areas that increase water retention time and sedimentation of solids. Dense vegetation, as found in riparian zones, can resist the movement and transport of water and associated materials (Cooper et al., 2000; Dabney et al., 2001). Increased water retention time can also promote other processes such as nutrient uptake by vegetation and microbes, adsorption to soil, and denitrification (van der Lee et al., 2004; Lohse et al., 2009; Pärn et al. 2012).

Besides natural processes, the agricultural activities influence the retention of sediment and nutrients in the valley bottom. Farming practices (hoeing, transplantation and weeding) can make the soil more susceptible to erosion (Labrière, 2015). Exceptionally high TSS and TP exports were observed in Tonga-2 and Gahenerezo reaches which are dominated by irrigated rice farming (Table 3-3). Gahenerezo is also densely populated and Tonga-2 has an aerodrome, which may add to increased exports. High TSS and TP retention were observed in Gihindamuyaga reach, exclusively composed of grass/forest, while high TN retention was observed in Rwasave reach dominated by ponds/reservoirs (Table 3-3). Settling of sediments is especially promoted when water retention time is increased in reservoirs and ponds (Rădoane and Rădoane, 2005). Rice farming appeared to have an intermediate position between standing water and other land covers: the flood water allows sediment to settle, but ploughing and the use of fertilizers (silage, manure, urea and various types of inorganic fertilizers) may create an export of sediments and nutrients. Standing water allows processing of dissolved nutrients, e.g. through uptake by phytoplankton or macrophytes. Ammonia sorption on soil substrate in wetlands (Kadlec and Wallace, 2009) may play a role in Tonga-1, where the unexpected high TN retention could be related to clay mining activities resulting in the export of ammonia sorbed to the clay material. Farming activities directly influence shallow subsurface flows, e.g. by disturbing anaerobic conditions and decreasing denitrification. In agricultural reaches, the use of fertilizers creates an additional nutrient input, while export in biomass (through vegetation or crop harvesting) contributes to reducing the transport of N and P.

During base flow, differences in nutrient yield among valley bottoms with different LULC were small. For sediment and P runoff, the impact of farming was highest during high flow situations. In many catchments, the major part of the annual phosphorus load is transported during storm flows following a period of base flow with high phosphorus build-up. In these periods, material accumulated during the dry season is exported at the onset of high rainfall (e.g. Demars et al., 2005). In this study, the highest exports were observed during the early stages of the vegetable growing season. The later stages of farming were characterized by small exports and sometimes retention. The peak of farming activities coincides with high flow events, and this is when valley bottoms are most vulnerable to the loss of sediment and nutrients. Impacts of farming are mostly related to disturbances of subsurface flows, soil (land preparation, ploughing, pond draining), and vegetation (weeding, harvesting) and to application of fertilizers. In contrast to sediment and P, flows of N are likely mostly affected by disturbance of the shallow groundwater processes and by increasing the retention time of water. Net yields of N were more evenly distributed throughout the year, except for the highly positive net yields measured in reaches with ponds/reservoirs during rainfall periods and related to water storage (Figure 3-4). Ponds and reservoirs retain TSS and TP during the water filling/operation period, but export them during periods of water release or when ponds are drained for fish harvesting or dredging (Figure 3-5).

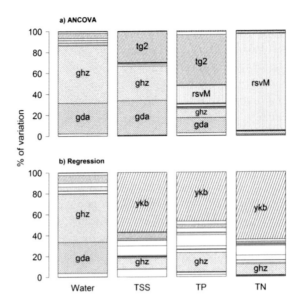

Figure 3-7. Variance contributions of different reaches in the two sets of models for a) ANCOVA models; and b) regression models. Names within the bars indicate the reaches representing the major part of the variance: Gihindamuyaga (gda), Gahenerezo (ghz), Tonga-2 (tg2), Rwasave (rsvM), and Nyakariba (ykb) (see also Figure 3-1).

Worldwide, natural land cover (grassland, forest, wetlands) is progressively declining while agricultural and urban areas are increasing (van Asselen *et al.*, 2013; WWF, 2014). Conversion of wetlands to agriculture as a strategy to increase food security is widespread throughout eastern and southern Africa (Schuyt, 2005; Wood *et al.*, 2013), including Rwanda (MINAGRI, 2009; 2010a,b). Mitigating such impacts requires the incorporation of land use and land cover types that store water and reduce nutrient losses. This could be done by protecting grass and forest vegetation in riparian zones and including ponds and reservoirs in the mix of land uses in valley bottoms. It is important to control the loss of sediment and nutrients at the critical stages of farming, notably during ploughing and harvesting. This can be done through the use of conservation agriculture techniques, such as ploughing with minimal soil disturbance, minimal use of water and fertilizers, use of retention ponds and wetland buffer zones to reduce water, sediment and nutrient transport (Drechsel *et al.*, 1996; Labrière *et al.*, 2015). Verhoeven *et al.* (2006) suggested that a certain proportion of wetland vegetation (2-7%) in riparian zones can significantly control the export from catchments of sediments and nutrients. LULC types with increased retention time can be placed in the riparian zone to intercept agricultural effluents. Management efforts should focus on spatial and temporal distribution of those factors (vulnerable and bare soil, application of manure and fertilisers, waste discharge, high rainfall) whose combination makes an area a critical source of sediment and nutrients to water bodies (Heathwaite *et al.*, 2000; Jennings *et al.*, 2003; Labrière *et al.*, 2015).

3.5 Conclusion

This study showed that the combination of discharge and agricultural land use created strong seasonal effects on the net yields of sediment and nutrients in valley bottoms in southern Rwanda. Natural land cover (grass/forest) and standing water (ponds/reservoirs) mainly retained sediments and nutrients, whereas farming activities (rice and vegetable farming) caused export of sediments and nutrients. Sediment and phosphorus transport were most affected by valley-bottom farming during high discharge periods which coincide with peak farming activity. Nitrogen transport was more evenly distributed throughout the year, and differences among land use types were smaller. The current policies in Rwanda of converting valley bottom wetlands to agriculture and increasing crop production pose a risk for further water quality deterioration unless appropriate management practices are considered. Sediment and nutrient losses can be controlled through the use of adaptive LULC (e.g. grass/forest and ponds) in the riparian zones to intercept the TSS, TP and TN from agricultural runoff, and by introducing conservation agriculture techniques that reduce the transport of sediment and nutrients at critical stages of farming.

3.6 Acknowledgements

This study was conducted in cooperation of the University of Rwanda (former National University of Rwanda) and IHE Institute for Water Education, Delft, The Netherlands. The research funds were provided by NUFFIC through an NFP (Netherlands Fellowship Programme) fellowship to the first author. We thank Andrew Jackson of Trinity College, Dublin, Ireland for advice on statistical analysis and Jochen Wenninger (IHE-Delft, The Netherlands) for his comments on hydrological aspects of the research.

4. Mesocosm studies on the effects of conversion of wetlands to rice and fish farming on water quality in valley bottoms of the Migina catchment, southern Rwanda[3]

Abstract

Agricultural development is critical for economic growth and food security. However, sediment and nutrient runoff generated by farming activities may constitute an important source of pollution and water quality deterioration. The objective of this study was to assess the effects of the conversion of wetlands to agriculture on the water quality in valley bottoms in the Migina catchment, southern Rwanda. Three valley bottom land use/land cover (LULC) types (fishponds, rice and wetland plots) were studied in a replicated mesocosm setup. LULC characteristics (hydrology, biomass growth) and farming practices (land preparation, water use, feed/fertilizer application) and the associated changes in total nitrogen (TN), total phosphorus (TP), total suspended solids (TSS), dissolved oxygen (DO), conductivity (EC), pH, and temperature were measured during two periods from 2011 to 2013. Results showed that fish farming used 7-8 times less water than rice farming (101 mm m^{-2} d^{-1} compared with 191 mm m^{-2} d^{-1}). Biomass increase in fish farming was much lower (30 g m^{-2} in 8-months) than in rice (2500 - 4500 g m^{-2} in 4 months) and wetland plots (1300 - 1600 g m^{-2} in 8-11 months). Over the two seasonal periods studied, higher concentrations of TSS, TP and TN in inflows and outflows were mainly associated with human activities (cleaning of water supply canals, rice plot ploughing, weeding and fertilizer application, and fishpond drainage and dredging). As a result, fishponds and rice plots generally had consistently higher TSS concentrations in surface outflows (5-9506 and 4-2088 mg/L for fishponds and rice plots, respectively) than in inflows (7-120 and 5-2040 mg/L for fishponds and rice plots, respectively). For TN and TP, results were more mixed, but with peaks associated with the periods of land ploughing, weeding and fertilizer application, fishpond drainage and dredging. In wetland plots, TSS and TN significantly decreased from the inlet to the outlet, owing to the absence of disturbances of the plots and probably other mechanisms (higher settling/adsorption, nutrient uptake and denitrification). Adoption of conservation farming techniques and efficient use of water and fertilizers would promote environmental protection and the sustainability of agricultural production. It is worth exploring the integration of fishponds for temporary storage of water, sediments and nutrients for reuse in crop farming, and of natural wetlands as buffer zones for sediments and nutrients from farming effluents during the critical periods of agricultural activities.

Keywords: Fish farming, rice farming, wetlands, nutrients, sediments, water quality

[3] Published as:
Uwimana, A., van Dam, A.A., Irvine, K., 2018. Effects of conversion of wetlands to rice and fish farming on water quality in valley bottoms of the Migina Catchment, southern Rwanda. Ecological Engineering 125, 76-86. https://doi.org/10.1016/j.ecoleng.2018.10.019

4.1. Introduction

Agriculture is vital for economic growth and food security, but also has an impact on ecosystem services and benefits (Davari et al., 2010). This is exemplified throughout eastern and southern Africa, where conversion of wetlands to agriculture is widespread (Rebelo et al., 2010; Asselen et al., 2013). In Rwanda, people historically congregated in hillside settlements, using valley-bottom wetlands as a source of water, wild foods, and medicines. With the increase in population during the 1980s, views on wetlands shifted, with more interest in their potential as fertile lands that can contribute to national food production, particularly for rice and fish farming. As rice can produce up to 7 tonnes per ha, much more than other crops, it was regarded as one of the most suitable crops in Rwandan inland valleys (MINAGRI, 2010b). Strategies to increase rice production by increasing the area under cultivation, and use of modern inputs such as seeds, fertilizers, pesticides and irrigation were implemented with vast areas of valley bottom wetlands converted into rice farms, and small dams and ponds constructed to support irrigation and fish production. Rice farming area increased from 3,549 ha in 2000 to 13,000 ha in 2009 and was expected to expand further to 28,500 ha by 2018 (MINAGRI, 2013a). Fertilizer application increased steadily, from a national average of about 4 kg ha^{-1} in 2006 to 30 kg ha^{-1} in 2013, and expected to reach 45 kg ha^{-1} in 2017/18 (MINAGRI, 2014). Similarly, aquaculture production of 4,038 mt in 2007 (Mwanja et al., 2011) was expected to increase to about 35,000 mt by 2020 (MINAGRI, 2011).

While these developments are positive for food security, their environmental impacts may be negative because of the alteration or destruction of wetland ecosystems and biodiversity, and associated impact on their regulating ecosystem services. This has led to changes in water quality, not only of Rwanda's streams, rivers and lakes, but of the Kagera river, the main tributary that flows into the transnational Lake Victoria. Depending on farming practices and management, sediment and nutrients from agricultural areas can contribute significantly to freshwater eutrophication (Jennings et al., 2003; Díaz et al., 2012). These are clear risks owing to the hilly topography of Rwanda and inappropriate soil conservation measures.

The retention function of wetland ecosystems for water, sediment and water chemistry is well documented (Verhoeven et al., 2006; Kadlec and Wallace 2009). With the increasing need for food security and economic development, wetlands have been converted into crop and fish farms which changes their retention function. The extent to which agricultural land use generates or retains sediments and nutrients depends on both the intrinsic characteristics of the farming systems, and on the specific practices related to each farming system. The characteristics of a farming system are the result of hydrological and physiological processes of the organisms grown (e.g. productivity, nutrient use efficiency), while the farming practices depend on many factors, including available technology and inputs, socioeconomic factors, and agro-climatological conditions. Fishponds can act seasonally as retention reservoirs for water, sediments, nitrogen and phosphorus that can subsequently be released and degrade the environment (Rahman and Yakupitiyage, 2003; Barszczewski and Kaca, 2012). Low recovery (5-6%) of nitrogen and phosphorus in fish biomass have been found in aquaculture ponds in Vietnam, with the major part of the nutrients accumulating in sediments or exported in fishpond effluents (Nhan *et al.*, 2008). Nutrient recovery in irrigated and fertilized rice farming rarely exceeds 40% (Schnier, 1995; Krupnik *et al.*, 2004), while water productivity for rice farming typically varied between 0.20 and 0.68 kg of grain per m^{-3} of water input (Bouman *et al.*, 2005; Christen and Jayewardene, 2005). The water requirement of fishponds, estimated at 10.3 m^3 kg^{-1} of fish production by Sharma *et al.* (2013) is even higher than crop farming. Natural wetland vegetation regulates hydrological processes, acting as both a sink and source of sediments and nutrients (Mitsch and Gosselink, 2000; Bullock and Acreman, 2003).

A catchment scale study in the Migina catchment, southern Rwanda (Uwimana *et al.*, 2017; Chapter 2) showed seasonal trends in water quality (build-up, washout and dilution) that were associated with changes in surface water flow and differences in land use. Nitrogen and phosphorus accumulating in the catchment during the dry season were washed into water courses during the early stages of the higher flows and diluted at the end of rainy periods. Generally, conductivity, temperature, dissolved oxygen and pH decreased with increasing discharge. Each of the studied land uses (rice farms, vegetable farms, ponds/reservoirs and grass/forest) shifted seasonally from being a source to a sink for nutrients and sediments, although overall rice and vegetable farming contributed more to the export than ponds/reservoir and grass/forest. However, it was not clear if water bodies or natural land cover are intrinsically more efficient in retaining nutrients and sediment than agricultural plots (e.g. by storing more nutrients in biomass or in soil), or if the observed differences were caused by farming practices (e.g. water management, or tillage practices).

This study was conducted to investigate the effects of land use characteristics and land management practices on water quality in experimental systems with controlled hydrology and farming practices. Mesocosm models of three types of land use & land cover (LULC) in valley bottom of southern Rwanda (fish farming in ponds, irrigated rice farms and natural wetland vegetation) were used to investigate water quality dynamics from the inlet to the outlet. The overall objective was to assess the effects of conversion of wetlands to fish and rice farming on water quality. First, we described the characteristics (hydrological processes, productivity) and practices (land preparation, feed or fertilizer application) in the three LULC types. Then, we compared inlet and outlet concentrations of total nitrogen, total phosphorus, total suspended solids, dissolved oxygen, electrical conductivity, pH, and temperature among the three LULC types. Finally, we assessed the effects of land use characteristics and management practices of the three LULC types on water quality in Rwandan valley bottom streams.

4.2. Material and Methods

4.2.1. Study area and period

This study was conducted at the Rwasave fish farming station of the University of Rwanda. The station (geographic coordinates 2° 40' S, 29° 45' E) is located 2 km east of Huye Town, in the Rwasave valley-bottom wetlands (Figure 4-1a), and connected to the headwaters of Migina River, a tributary of the Akagera and Nile rivers. The climate is temperate tropical humid with a pronounced dry period between June and September and low thermal amplitude. The average temperature is around 20° C and the annual rainfall varies around 1200 mm (SHER 2003). The climate in the area is mainly influenced by the proximity to the equator at 2° to the north, Lake Victoria in the East, and the altitude (1625 m). Rwasave fish farming station was established in 1982 and consists of 103 fishponds of different sizes (Figure 4-1b). The station produces Nile tilapia (*Oreochromis niloticus*) and African catfish (*Clarias gariepinus*). Water is supplied by the Rwasave River, which flows from the headwaters of Migina catchment (Munyazi sub-catchment). Plots with rice and wetland macrophytes were established at the fish farm (Figure 4-1b).

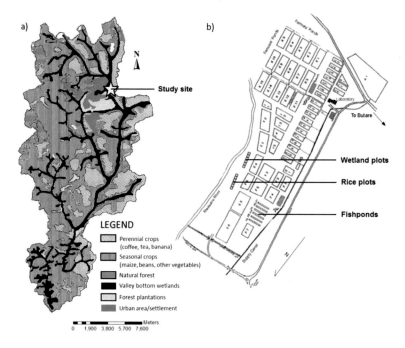

Figure 4-1. a) Migina catchment with land use (source: Uwimana et al 2017), and b) Rwasave fishpond station layout with experimental study plots (source: adapted from PD/ACRSP 1999).

4.2.2. Experimental setup

Six replicates for each LULC type (fishponds, rice and wetland plots) were used. Ponds and rice plots were operated using the common farming practices in southern Rwanda (Figure 4-2). The fishponds were operated as stagnant systems with limited outflow while rice and wetland plots were operated with continuous inflow and outflow. Water entered and left the experimental plots through pipes where water flow measurement was feasible. The study covered two periods: from January to December 2011 (Period 1); and from September 2012 to July 2013 (Period 2), with slight variations in experimental periods among the three systems within these periods.

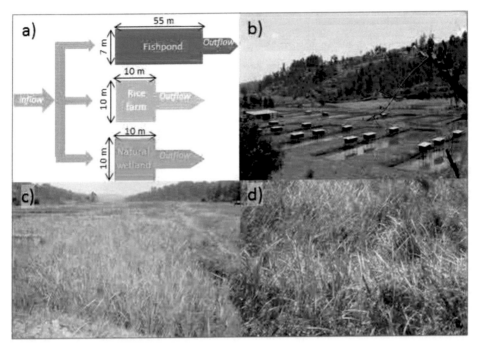

Figure 4-2. Experimental design with three land use/land cover (LULC) types. a) Schematic diagram of experimental setup; and photos of b) fishponds; c) rice plots; and d) wetland plots.

Each fishpond was approximately 55 m long, 7 m wide with depth varying from 1 to 1.5 m (Figure 4-2b). Maximum water depth was 1.05 m while the minimum was 0.42 m. Five of the six ponds were stocked with Nile tilapia (*Oreochromis niloticus*) of 23 g (Period 1) and 9 g (Period 2) mean individual fresh weight at a density of 2 fish m^{-2} (840 fish per pond of 385 m^2). Rice bran was used to feed the fish at a rate of 2.5 g fish^{-1} day^{-1}. Feeding was reduced after the third month to avoid overfeeding and water quality problems. The bottom and sides of the ponds were lined with clay to minimize water seepage and water was intermittently supplied (once or twice a month) to compensate for evaporation losses and keep the ponds filled.

Six replicate rice plots of 10 m long, 10 m wide and 0.4 m deep were subjected to the same land preparation, irrigation water supply, rice planting, weeding, fertilizer and herbicide application (Figure 4-2c). The plots were enclosed within 40 cm high clay peripheral dikes to prevent uncontrolled surface water flows. With the exception of the planting, fertilizer application and harvesting periods, the plots were supplied with water to keep them flooded (0.15 m), which also controlled weed growth. Hydrological and water chemistry variables in the inflow and outflows were monitored throughout the experimental period. Rice was grown at a density of 50 seedlings per m^2 and 3 kg of mineral fertilizer (2.5 kg of NPK and 0.5 kg of urea per plot) was applied following local farming practices. Basal application was done for NPK during transplantation of seedlings while urea was spread over the plots soon after weeding.

Six wetland plots with the same size as the rice plots were also established (Figure 4-2c). These plots were not subject to any management practices, except that water was supplied continuously to also maintain a depth of 0.15 m. As a result of this, a dense stand of the common emergent wetland plant *Cyperus latifolius* developed.

A check list of activities (land preparation, stocking/planting, feed/fertilizer application, harvesting and dredging) taking place in each LULC was kept during the experiment. The quantities of feed and fertilizer applied for fish and rice farming, respectively, were recorded before every application.

4.2.3. Data collection

4.2.3.1. Hydrology

Water flows were quantified in terms of inflows (surface inflow and rainfall), outflows (surface outflow and evaporation) and groundwater exchange during Period 2 from September 2012 to May 2013. Surface inflow and outflow were measured at inlet and outlet pipes by measuring the time required for a container of a known volume to be filled with water. Rainfall was recorded using a rain gauge installed at the study site. Evaporation was measured as pan evaporation. Evapotranspiration was estimated based on evaporation and a crop coefficient depending on the four growth stages of rice (initial, development, mid-season and late season with 30, 30, 80 and 40 days and crop coefficients of 1.05, 1.20, 0.90-0.60 and 1.00 respectively; FAO 1998). For the mid-season stage, 0.75 as the average was used. The crop coefficient for wetland vegetation was assumed to be 0.95 (Drexler *et al.*, 2008). Net groundwater exchange (So - Si) was calculated based on the water balance (1).

$$S_o - S_i = P + Q_i + R - (E + Q_o + \Delta H)$$

in which S_i and S_o are, respectively, seepage in and out, P is precipitation, Q_i and Q_o are, respectively, inflow and outflow, R is runoff, E is evaporation, and ΔH is change in storage. Positive values of S_o-S_i indicate seepage out of, and negative values seepage into, the plot. In rice and wetland plots, storage (ΔH) was assumed zero because the plots were kept flooded with water throughout the experimental period. In fishponds, ΔH was determined as the difference between water level recorded in the morning (d_1) and evening (d_2). The high clay peripheral dikes prevented uncontrolled surface water flows. R was assumed to be zero.

4.2.3.2. Biomass

Fish, rice plant and grass samples were taken at the start of the experiment and periodically until harvest. Each month, forty fish from each pond were taken, weighed (g) and measured (cm) and returned to the pond. For rice and wetland plots, rice and grass belowground and aboveground biomass were taken in three quadrats of 1 m^2 in each experimental plot at the start, and then monthly until the end of the experiment. All fish and plant samples were dried in an oven at 65 °C for 48 hours and weighed. The biomass growth (g dry matter m^{-2} d^{-1}) was calculated from start to end of the experiment (274 days and 259 days in Periods 1 and 2, respectively for fish; 150 and 127 days in Period 1 and 170 and 191 days in Period 2 for rice; and 345 days and 274 days for Periods 1 and 2 for wetlands).

4.2.3.3. Water quality dynamics from the inlet to the outlet

Water quality was measured monthly at the inlet and outlet of the experimental plots. Temperature (°C), pH and EC (μS cm^{-1}) were measured on-site using a Portable Multimeter 18.50.SA (Eijkelkamp, Giesbeek, The Netherlands). EC was temperature corrected (specific conductance, SC) at 25 °C with a temperature correction coefficient of 0.0191 as directed in Standard Methods for the Examination of Water and Wastewater (APHA, 2005). Dissolved oxygen (DO) was measured using a Portable DO Meter Accumet Waterproof AP74 (Fisher Scientific, Pittsburgh, USA). Total suspended solids (TSS) was determined on-site using a Portable colorimeter DR/890 (HACH, Colorado, USA). Samples for total phosphorus (TP) and nitrogen (TN) were preserved by acidification to pH = 1-2, using sulphuric acid (0.1 N) and transported to the laboratory of the University of Rwanda (Huye campus) to determine TN and TP using the persulfate digestion method (APHA 2005).

4.2.3.4. Data analysis

Water flows were summed per month to compare the water balance of the three LULC systems. Linear regressions were calculated among the flow components. Non-parametric pairwise t-tests (Mann-Whitney test) were used to compare water quality in inlets and outlets of the three LULC systems. To compare the three systems, the repeated water quality measurements were analysed using the approach for analysis of longitudinal data presented by Everitt and Hothorn (2010). Linear mixed effects models were formulated for Periods 1 and 2 separately, using land use type and month as fixed variables and individual plots as a random effect to account for the dependence of the repeated measurements. Both random intercept and random slope models were used, and the model with the lowest AIC was selected. Mean water quality in the land use types for each period was compared using least square means with Tukey adjustment. All calculations were done using function wilcox.test for the pairwise t-test, lme for estimating the mixed models, function lsmeans for posthoc analysis, and function lm for regression, all part of R software version 3.3.1 (R Core Team, 2016).

4.3. Results

4.3.1. Land use practices

In rice farming, land preparation (hoeing and levelling) started with the onset of the rainy season in August-September or in February, followed by transplanting of seedlings, water supply and fertilizer application (Figure 4-3). Water supply continued to the period preceding the harvest, while weeding, and urea and pesticides application were done at the end of the second and fourth month. Harvesting took place in December-January (short dry period) or June-July (dry period). Fish farming started in September with filling the ponds with water and stocking the fish, followed by daily feeding and intermittent water refilling. The culture period was concluded with lowering of the water level by draining, fish harvesting and pond dredging. Dredging was done at the beginning or/and at the end of the fish growing season. This practice generates a large amount of accumulated substances like pond sediment and probably nutrients. For the wetland plots, water application was the only management practice.

4.3.2. Hydrological processes

The water balance varied according to land use type (Figure 4-4) and farming stages. Total monthly rainfall in the period September 2012-May 2013 ranged from 97 mm in September to 360 mm in April (Figure 4-4a). For monthly water balance calculations, rainfall measurements on 16, 17, and 18 days were used to estimate water balance for, respectively, wetland, fishpond and rice plots, which coincided with measuring other hydrological flows (inflow, outflow, water level change, evaporation). This resulted in some differences in rainfall numbers across LULC plots.

In fishponds, filling the ponds in September (786 mm, or 98% of all inflows) was by far the greatest flow compared to all other flows throughout the season (Figure 4-4b). Refilling was intermittent and was the highest in February (46 mm) when there was no rain for two weeks. The lowest inflow was observed in April when rainfall was highest. Net seepage into the ponds occurred in February (12 mm) and May (52 mm), constituting 44% and 20% of all inflows in those months, respectively. Fishpond surface outflow varied from 0 to 688 mm mo^{-1}, with the highest observed in May at the end of the fish farming season during pond emptying and dredging. No surface outflows were observed in October, November and December, which reflects the stagnant character of the ponds. The surface outflows observed in September, January, February, March and April were caused by the lowering of the water for fish sampling and water renewal. Excluding the period of pond filling and the periods without rain (in September and February, respectively), rainfall was the most important supply flow for the fishponds. Evaporation constituted the most important water loss (49 mm or 72% of losses during February, the period without rain). Net groundwater exchange was dominated by seepage out and was highest in April (rainiest month) with 157 mm (60% of losses).

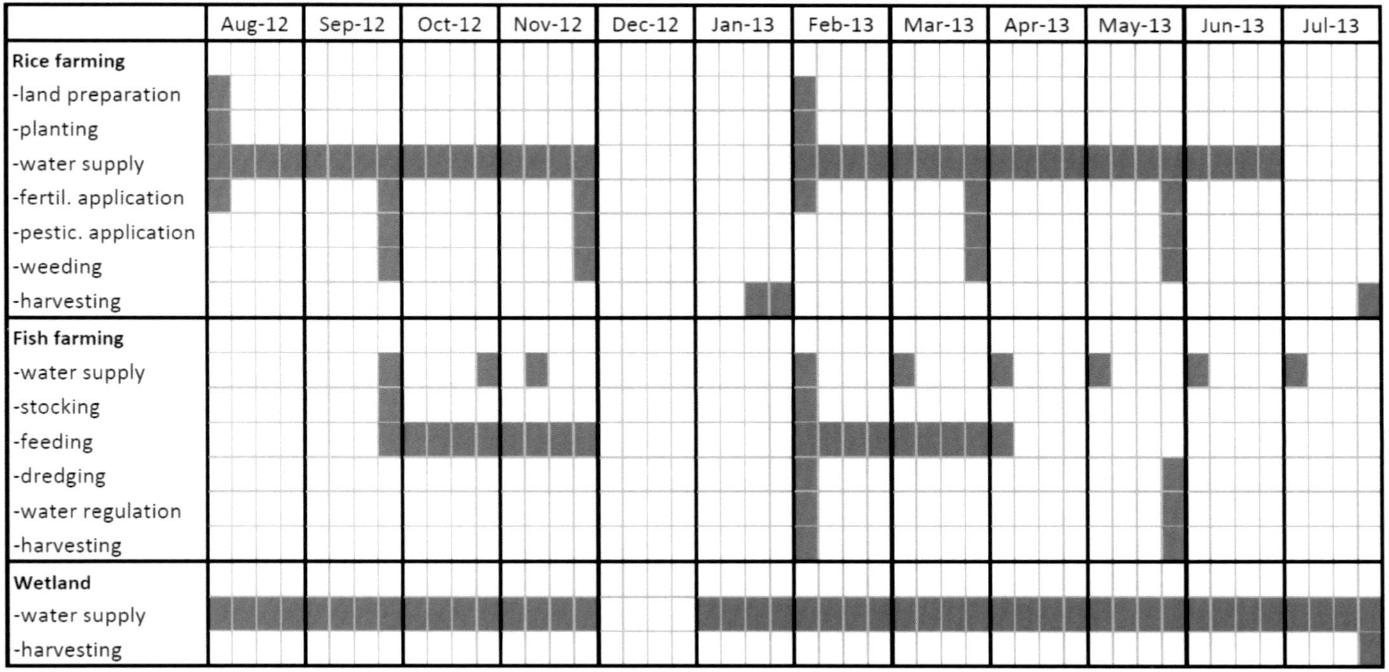

	Aug-12	Sep-12	Oct-12	Nov-12	Dec-12	Jan-13	Feb-13	Mar-13	Apr-13	May-13	Jun-13	Jul-13
Rice farming												
-land preparation	■						■					
-planting		■					■					
-water supply	■	■	■	■			■	■	■	■	■	
-fertil. application		■	■	■				■	■	■		
-pestic. application												
-weeding												
-harvesting						■						■
Fish farming												
-water supply		■	■	■			■	■	■	■	■	■
-stocking				■				■		■		
-feeding			■	■			■	■	■			
-dredging										■		
-water regulation										■		
-harvesting												
Wetland												
-water supply	■	■	■	■	■	■	■	■	■	■	■	■
-harvesting												■

Figure 4-3. Land , water and agricultural management practices for the three LULC types in Period 2 (August 2012 - July 2013). Management practices were similar in Period 1.

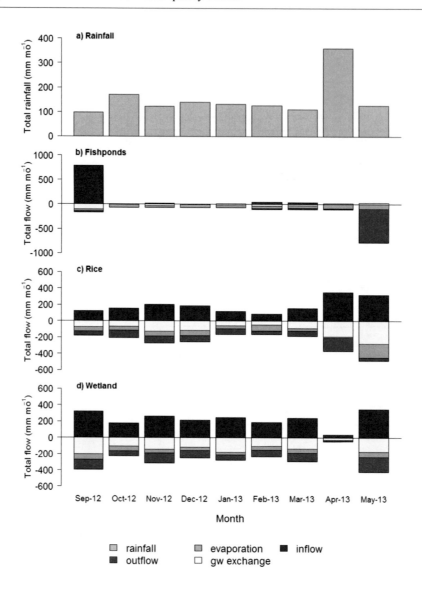

Figure 4-4. Water flows in experimental fishponds, rice and wetland plots in Period 2 (all in mm mo⁻¹).

Compared with fishponds, flows in rice plots were less variable with seasons (Figure 4-4c). Surface inflows varied from 89 to 356 mm mo^{-1} and were higher than rainfall (51 to 90% of total inflow). Higher inflows were observed in April and May as a result of higher flows in the supply canals that pushed more water into the rice plots. Lower values were observed in September, January and February during land preparation and harvesting. In contrast with fishponds, where seepage inflow was observed in some periods, the groundwater exchange consisted entirely of seepage out and varied from 54 to 241 mm mo^{-1}. Lower values were observed in January, February and March as a result of lower inflows (lower water supply and rainfall), while higher values were observed in April and May as a result of higher water supply and rainfall. Seepage out (144 mm mo^{-1}) constituted about half of the water gain (281 mm mo^{-1}, with 191 mm mo^{-1} of surface water inflow and 90 mm mo^{-1} of rain). Evapotranspiration and surface outflows were smaller with of 61 mm mo^{-1} and 76 mm mo^{-1}, respectively.

In the wetland plots, surface water supply (187-350 mm mo^{-1}) exceeded rainfall (0-139 mm mo^{-1}), while there was no seepage into the plots (Figure 4-4d). Seepage out varied from 61 to 254 mm mo^{-1} and was greater than surface outflow (12-179 mm mo^{-1}) and evapotranspiration (13-70 mm mo^{-1}). Higher seepage out was associated with higher inflows, while less seepage and surface outflow were associated with low inflow and rainfall.

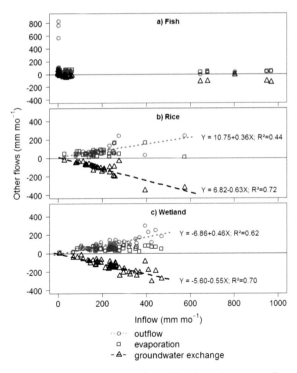

Figure 4-5. Relationship between water inflows (X-axis) and water outflow, evaporation/ evapotranspiration and groundwater exchange in three LULC-types (a: fishponds; b: rice plots; c: wetland plots) during Period 2.

Fishponds were supplied with half as much water (101 mm mo^{-1}) as the rice (191 mm mo^{-1}) and wetland plots (230 mm mo^{-1}), probably as a result of the clay lining that reduced water seepage, while rice and wetland plots were irrigated continuously. Seepage into the ponds was only observed in February (no rain) and May (water level lowering and pond emptying). Seepage outflow from fishponds (53 mm mo^{-1}) was about three time less than in rice and wetland plots (144 and 153 mm mo^{-1}, respectively). Rainfall (70-95 mm mo^{-1}) contributed 65 % of the water flow into fishponds, but much less in rice (35 %) and wetland plots (25 %). Evaporation in fishponds contributed more to water loss (41 %) than in rice (23 %) and wetland plots (20 %). With the exception of the pond dredging occurring at the end of the farming season, fishpond surface outflow was low or non-existent (0 - 20 mm mo^{-1}) throughout the farming season compared with rice (42-159 mm mo^{-1}) and wetland plots (12-179 mm mo^{-1}). In rice and wetland plots, both outflow and water loss to the groundwater were strongly related with inflow (significant regressions, with R^2 ranging from 0.44 to 0.72) whereas in fishponds no relationship was evident (Figure 4-5).

4.3.3. Biomass production

Biomass change with time varied strongly with land use types (Figure 4-5). Fish biomass in both 8-month periods increased to about 30 g m^{-2} fresh weight, while rice biomass increased to about 4500 g m^{-2} fresh weight in 4 months (except in the second crop of Period 1, when it reached only 2500 g m^{-2}). Biomass in the wetland plots reached about 1300 g m^{-2} in Period 1 and 1600 g m^{-2} in Period 2 (Figure 4-6).

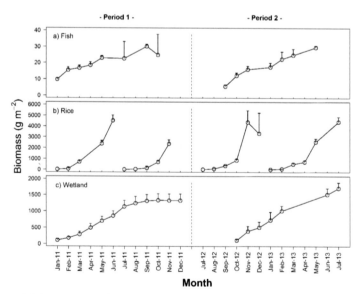

Figure 4-6. Biomass growth (fish, rice plants and wetland grass) in three experimental LULC systems (fishponds, rice and wetland plots) during two experimental periods. Data represent means of 5 (fishponds) and 6 (rice and wetland) plots ± standard deviation.

4.3.4. Water quality

Figure 4-7 and Table 4-1 show the differences between inflow and outflow water quality parameters. Water temperature was generally higher in the outflow from fishponds and rice fields, but lower in the outflow of wetland plots compared with inflow. Water temperature from wetland plots was generally about 2 °C lower in the outflow than in the inflow (Table 4-1). For pH, differences between inflow and outflow were small on average, all within 0.25 pH units from zero. In wetlands, pH in the outflow was consistently lower than in the outflow by about 0.2 units. In fishponds, pH was generally lower in the outflow in Period 1, but higher in the outflow during a large part of Period 2, except at harvest. Differences in EC among inflow and outflow were very small and not significantly different among LULC types. DO concentrations were lower in the outflow than in the inflow of wetland plots by about 1.2 mg L^{-1} in both periods. In rice plots, differences in DO were very small, whereas in fishponds there were differences between Periods 1 and 2 again, with lower DO in the outflow in a part of Period 1 and mostly higher DO in the outflows during Period 2, except on the day of harvesting (Figure 4-7).

Differences between outflow and inflow concentrations of TSS, TP and TN in fishponds were most striking on the last days of Periods 1 and 2. Exceptionally high TSS concentrations (1169-9605 mg L^{-1}) were observed in fishpond effluent in May during pond dredging at the end of the farming season. In rice plots, TSS concentrations were much higher in the outflow in some months, but small in other months. High TSS concentrations (100-2088 mg L^{-1}) were observed in August and January (land hoeing, levelling, planting) and throughout the farming seasons (November, December, March) during weeding and fertilizer application. Overall, and leaving the TSS results in fishponds in the last sampling month out of consideration, differences among LULC types in inflow and outflow sediment and nutrient concentrations were not significant, except for TSS in wetlands which showed significantly lower TSS in the outflow (lower on average by 10 mg L^{-1} in Period 1 and 106 mg L^{-2} in Period 2) compared with rice and fishponds (Table 4-1).

4.4. Discussion

In this study we assessed the effect of conversion of natural wetlands into rice and fish farming. Rice plots and fishponds were selected because they are very common and promising land uses that can boost agricultural development and food security. Wetland plots represent the original valley-bottom land cover and may be used to naturally improve sediment and nutrient retention capacity. Fishponds operate mainly as stagnant systems, with the major part of inflow at the beginning of the season (start-up period) and whenever there is a need to compensate losses to evaporation and seepage, while outflow happens mainly at the end of the season when ponds are drained and dredged, and fish harvested. Exchange with the groundwater is minor. On the other hand, rice and wetland plots received water continuously to keep the plots flooded, while compensating for evapotranspiration and seepage. This resulted in constant outflows of water through surface flows and seepage to the groundwater. Biomass accumulation was highest in rice plots and lowest in fishponds, with wetland plots in between. Wetland plots lowered the temperature and the DO concentration of the flood water. Wetlands also lowered, but rice plots and fishponds raised the TSS concentration of the floodwater. The impact on nutrients in the outflow water was small and differences among LULC types were not significant. The biggest impacts on sediment and nutrient outflow were related to key periods of land preparation (ploughing), crop management (weeding, fertilizer application) and harvesting (pond draining, dredging).

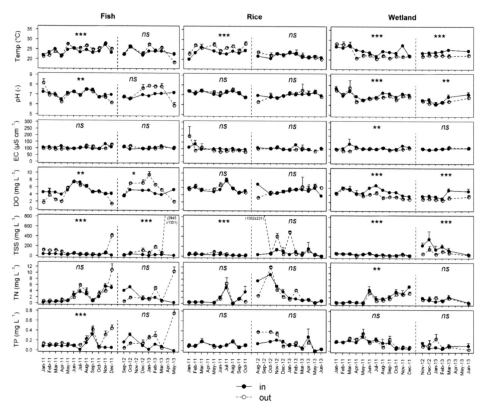

Figure 4-7. Water quality in inflow and outflow of experimental plots for three land use types (fishponds, rice plots and wetland plots) in experimental Period 1 (January - December 2011) and Period 2 (August 2012 - July 2013). Data are means of 5-6 replicate plots ± standard error. Stars indicate the outcome of paired non-parametric t-tests for the difference between inlet and outlet values for each period and land use type (Mann-Whitney test; *** p<0.001; ** p<0.01; * p<0.05; ns = not significant).

Table 4-1. *Mean (least square means ± standard error) differences in temperature (Temp), pH, electrical conductivity (EC), dissolved oxygen (DO), total suspended solids (TSS), total nitrogen (TN) and total phosphorus (TP) concentrations (inflow minus outflow) for three LULC types and two experimental periods (Period 1: January-December 2011; Period 2: Aug 2012 - June 2013). Negative numbers indicate higher outflow than inflow parameters. Within periods, parameters sharing the same superscript are not significantly different from the other land use categories (linear mixed effects model, and post-hoc pairwise comparison test, P< 0.05). For TSS, TN and TP the measurements in the last month of both periods in fishponds, and of the first month of Period 2 in rice plots were excluded in the analysis. For more explanation, see text.*

Period	Land use	Temp (°C)	pH (-)	EC (µS cm⁻¹)	DO (mg L⁻¹)	TSS (mg L⁻¹)	TN (mg L⁻¹)	TP (mg L⁻¹)
1	fishpond	-0.98 ± 0.32^b	0.12 ± 0.050^{ab}	-0.27 ± 5.49^a	0.71 ± 0.18^b	-25.11 ± 4.44^a	-0.23 ± 0.31^a	-0.22 ± 0.018^a
	Rice	-2.37 ± 0.29^a	-0.041 ± 0.056^a	14.43 ± 6.41^a	-0.17 ± 0.20^a	-18.46 ± 4.82^a	0.37 ± 0.33^a	0.38 ± 0.019^a
	wetland	1.73 ± 0.29^c	0.16 ± 0.051^b	4.98 ± 5.53^a	1.23 ± 0.18^b	10.26 ± 4.41^b	0.62 ± 0.31^a	0.60 ± 0.018^a
2	fishpond	-0.42 ± 0.35^a	-0.17 ± 0.12^a	-4.93 ± 3.04^a	-1.38 ± 0.32^a	-51.74 ± 23.71^a	1.13 ± 0.29^a	1.27 ± 0.326^b
	Rice	-0.24 ± 0.37^a	0.10 ± 0.12^a	2.43 ± 3.33^a	-0.023 ± 0.33^b	-68.50 ± 25.13^a	0.084 ± 0.31^a	-0.025 ± 0.386^a
	wetland	2.21 ± 0.40^b	0.23 ± 0.13^a	-1.73 ± 3.25^a	1.17 ± 0.35^b	106.08 ± 25.25^b	0.099 ± 0.31^a	-0.0098 ± 0.343^a

In wetland plots, water temperature, pH, DO and TSS consistently decreased from the inlet to the outlet. This can be attributed to the dense vegetation that shades the ground and intercepts sediment from the incoming water. Despite comparable hydrological flows and biomass in rice and wetland plots, the absence of human activities in the natural wetland makes its outflow less concentrated in sediments and nutrients. Another key difference with rice plots is that no nutrients were added to the wetland plots, likely resulting in smaller flows of sediments and nutrients to the outflow and groundwater of wetland plots. Lower TN concentrations in the effluent of the wetland plots can be explained by higher settling/adsorption, nutrient uptake and denitrification (Ciria et al., 2005; Kadlec and Wallace 2009). While nitrogen can undergo complex transformations involving particulate settlement, uptake into biomass and possibly denitrification (Ciria et al., 2005), phosphorus can settle, be absorbed or chemically bound to soil, especially the ferrallitic soil that is typical of the study area (Schulz et al., 2003).

Although the differences in internal processes among the LULC types can explain some of the water quality differences, it could be observed that TSS, TN and TP were higher in the outflow of fishponds and rice plots only during human activities disturbing the soil, such as hoeing, levelling, transplantation, weeding and fertilizer application in rice plots and fishpond dredging and drainage. These critical periods of farming were also observed by Uwimana et al. (2018; see Chapter 3) who observed higher sediment and nutrient losses at the early stages of farming and during high flows in farms in the Migina catchment, while the later stages of farming were characterized by small exports and sometimes retention. The loss of sediments and nutrients from the farmed plots through surface flows is determined mostly by the farming activities during high-flow periods, when sediments and associated nutrients are loose and become easily suspended and transported in the outflow. This runoff during farming activities could be reduced by limiting water flow to rice farms to the minimum needed to flood the plots and control weeds, thus minimizing water and nutrient losses to the groundwater.

More sustainable farming methods would also be an adequate response to the conversion of wetlands to agriculture. Large areas of wetlands in Migina catchment and similar catchments in Africa have been converted into farming areas for rice, vegetables, fish, sugarcane, groundnut, fruits and for livestock grazing (Schuyt, 2005; Wood et al., 2013). In response to the need to increase food security and economic development, more wetland conversion is expected in the future (MINAGRI, 2009; 2010a,b; FAO, 2011a). Wetland conversion combined with the expected more intense and frequent droughts and floods caused by climate change (FAO, 2011a) will result in higher losses of sediment and possibly nutrients (Hecky et al., 2003). Adopting conservation farming techniques would reduce soil disturbance (minimal tillage), maintain permanent soil cover (intercropping and agroforestry) during rainy seasons and reduce irrigation/drainage water losses (increase water use efficiency). Crops need appropriate rates of fertilizers, applied at the right time (period of vegetative growth and low water supply) to avoid fertilizer wastage through soil immobilization, emission, leaching, effluent discharge and erosion. Precision agriculture, applying water, nutrients and pesticides only in the right place and time (Day 2008) would promote environmental protection and the sustainability of agricultural production.

Considering the characteristics and management requirements of different LULC types, it is worth considering combinations of these culture systems. Although fishponds have a low nutrient use efficiency with respect to fish production (Nhan, *et al.* 2008; Kufel 2012), they temporarily trap water, sediments and nutrients that can be utilized at the end of the season. The effluent from fishponds during fish stock control and harvesting, water renewal and dredging could be contained in nearby ponds for agricultural reuse (e.g. rice farming). Water from the ponds can be reused for irrigation, thus promoting overall productivity and more cost-effective nutrient cycling (Kipkemboi *et al.,* 2010). Rice farming with minimum tillage, optimum use of water and fertilizers would reduce water, sediments and nutrient transport to water bodies. In contrast to low-input farming that exhausts soil and has a negative nutrient budget (Tittonell and Giller, 2013), irrigated and high-input rice farms disturb soil and sediments and have low nutrient recovery and low water use efficiency (Schnier, 1995; Krupnik *et al.*, 2004). More optimal nutrient management can be achieved when natural wetlands can be integrated with fish and rice farming and act as buffer zones for sediments and nutrients from farming effluents during the critical periods of ploughing, weeding, manure and fertilizer application and water releases from fishponds.

4.5. Conclusion

Fishponds operated mainly as stagnant systems with irregular water filling and drainage, while rice and wetland plots were kept flooded with continuous inflow and outflow. Water losses from fishponds were small and consisted mostly of evaporation during the culture period and drainage at harvest. In rice and wetland plots, water losses were much larger and consisted of seepage, outflow and evapotranspiration. In the three types of land uses, water quality differences between inflow and outflow were largest in wetland plots, with significantly lower nutrient and sediment concentrations in the outflow. In rice plots and fishponds, water quality differences were largest at times of farming activities which disturbed soil and water flows. The effluent from fishponds during fish stock control, water renewal and dredging could be contained in ponds for agricultural reuse, e.g. by growing rotational rice crops in fishpond sediment. Within valley bottoms, natural wetlands could be integrated with fish and rice farming and act as buffer zones for sediments and nutrients from farming effluents in critical periods of land ploughing, weeding, application of manure and fertilizers, and water release from fishponds.

4.6 Acknowledgements

This study was conducted in cooperation of the University of Rwanda (former National University of Rwanda) and IHE Institute for Water Education, Delft, The Netherlands. The research funds were provided by NUFFIC through an NFP (Netherlands Fellowship Programme) fellowship to the first author.

5. Mesocosm studies on the impacts of conversion of wetlands into fish and rice farming on sediment and nutrient loads to surface water[4]

Abstract

Alteration of water and nutrient cycles threatens land, water, climate, biodiversity and human health and livelihoods. The main objective of this study was to assess the effects of conversion of natural valley-bottom wetlands in southern Rwanda to rice and fish farming on sediment and nutrient flows. To achieve this, small scale mesocosm experimental wetland and rice plots and fishponds were established to measure sediment, nitrogen (N) and phosphorus (P) flows Sediment, N, and P discharges (concentrations and loads) in outflows were lower in wetlands than in fishponds and rice plots, indicating that wetlands retained and did not release sediment and nutrients. While wetland sediment retention was linked to the dense vegetation cover, low sediment, N and P releases were also attributed to low wetland disturbance, greater N and P transfer to the underground, higher uptake by plants and settling in sediments. Ploughing and weeding during the first three months of rice farming, and water renewal and dredging in the middle and end period of fish farming generate and discharge large amounts of sediment, N and P in the outflow of these systems. This makes fishponds a temporal sediment and nutrient storage during early farming stages and a source towards the end of farming. In contrast, rice farming generates sediments and nutrients early during the farming period (ploughing, weeding, transplantation and fertilizer application) and traps them towards the end. Despite higher fertilizer input in rice farms, N and P storage in soil decreased in rice farms (by 4.7 and 1.4%, respectively), but increased in fishponds (by 3.3% and 4.4%) and wetlands (3.8% and 1%). The decrease in nutrient soil storage was attributed to higher N and P uptake in rice plots (on average 662 and 270 mg $m^{-2}d^{-1}$ of N and P, respectively) than in wetlands (359 and 121 mg m^{-2} d^{-1} of N and P) and fishponds (7.4 and 4.4 mg m^{-2} d^{-1} of N and P). The increase in nutrient storage increase in wetlands was attributed to senescence and decomposition of vegetation and subsequent storage in wetland sediments. Efficient reuse and recycling of water, sediments and nutrients within catchments may be achieved through rotational rice-fish and wetland-rice farming to balance the sources and sinks of sediment and nutrients in the landscape.

Keywords: Fish farming, rice farming, wetlands, nutrients and sediments pathways

---

[4] Submitted to: Frontiers in Environmental Science

5.1 Introduction

Life on Earth is closely regulated by the availability and cycling of water and nutrients (MEA, 2005). Natural ecosystems like wetlands and forests are strong regulators of water and nutrients at local and global scales. They play a big role in controlling sediment transport to river systems. The increased water residence time and the presence of vegetation in wetlands lead to settling and trapping of sediments (Verhoeven *et al.* 2006; Reddy and Delaune 2008; Kadlec and Wallace 2009). Wetlands can receive water, sediment and nutrients, temporarily store them in water, sediments or in living matter, transform them into other chemical forms, and progressively release them again (de Groot 2005; Verhoeven *et al.* 2006; Reddy and Delaune 2008; Kadlec and Wallace 2009). Wetlands can thus function as sinks, transformers or sources of nutrients (Mitsch and Gosselink 2000).

Human manipulation of ecosystem properties to increase provisioning goods and services has greatly interfered with the normal pathways and rates of water and nutrient cycling that maintain healthy ecosystems. This has resulted in deterioration of livelihoods, climate, land and water through erosion, water pollution, eutrophication and sedimentation of aquatic systems (MEA, 2005). Conversion of natural wetlands to farming or other forms of human use can lead to impairment of their natural functions and cause elevated losses of sediment and nutrients from catchments (Ramsar Convention on Wetlands 2018). In Sub-Saharan Africa, an estimated 65% of natural wetlands occurring in Chad, Congo, Niger and Nile catchments are threatened by farming activities (McCartney *et al.* 2010; Davidson 2014). Many studies have shown the broad effects of farming on water quality (Hecky *et al.* 2003; Bagalwa 2006; de Villiers and Thiart 2007; Verhoeven *et al.* 2006; Uwimana *et al.* 2018a,b), but not much is known in detail about the effects of conversion of wetlands into farming on water and nutrient pathways and cycling processes. It is important to understand the effect of conversion of wetlands on sediment and nutrient flows to achieve a balance between the benefits of farming and the ecosystem services provided by natural wetlands.

In Chapter 4, we assessed the effects of conversion of wetlands to fish and rice farming on water quality (Uwimana *et al.* 2018b). Results showed that human activities like cleaning of canals, ploughing, weeding, fertilizer application, fishpond drainage and dredging were responsible for higher suspended solids, phosphorus and nitrogen concentrations in inflows and outflows of rice and fish farming. In contrast to fishponds and rice farms, suspended solids and nitrogen significantly decreased from inlet to outlet of natural wetland plots, mostly as a result of less human disturbance compared to the rice and fish systems. Moreover, fish farming used remarkably smaller amounts of water, but needed considerable amounts of feed and produced much lower biomass compared to rice farms and natural wetlands. However, in the study not all sediment and nutrient flows were fully covered. To understand the contribution and extent of each flow in the retention of sediments and nutrients, it is important to look at all sediment and nutrient pathways.

The overall objective of this study is to assess the effects of conversion of natural wetlands into rice and fish farming on sediment and nutrient pathways. To achieve this, small-scale experimental wetland and rice plots and fishponds were established. The hydrological flows for fishponds, natural wetlands, and rice farms were measured, along with the sediment, nitrogen and phosphorus content in these flows. Then we quantified and compared the sediment, nitrogen and phosphorus loads and retention among the three systems.

5.2. Material and Methods

5.2.1 Study area and period

The study was conducted at the Rwasave fish farming station of the University of Rwanda. The station is located 2 km eastward of Huye Town, in the valley-bottom of the Rwasave wetlands that are connected to the headwaters of Migina River, a tributary of the Akagera River (geographic coordinates 2° 40' S, 29° 45' E; Figure 1a). The climate is temperate tropical humid with a pronounced dry period between June and September and low thermal amplitude. The average temperature is around 20°C and the annual rainfall varies around 1200 mm. The climate in the area is mainly influenced by the proximity to the equator at 2° to the north, Lake Victoria in the East, and the altitude (1625 m). The station was established in 1982 and consists of 103 fishponds of different sizes (Figure 5-1b) and produces 2 types of fish: Nile tilapia (*Oreochromis niloticus*) and African catfish (*Clarias gariepinus*). Water is supplied through Rwasave River from the headwaters of Migina catchment. Rice farms and plots of wetland macrophytes were established on the station to be used as small-scale experimental land use/land cover (LULC) units (Figure 5-1b).

5.2.2 Experimental setup

The study was conducted in the period September 2012 to July 2013 and used five ponds (55 m long, 7 m wide and depth varying from 1 to 1.5 m), six rice plots (10 m long, 10 m wide and 0.4 m deep) and six plots with native wetland macrophytes (10 m long, 10 m wide and 0.4 m deep). Fish and rice farming mimicked the common farming practices in the area. Fishponds were filled with water at the start of the experiment, stocked with Nile tilapia (*Oreochromis niloticus*) at 2 fish m^{-2} (840 fish per pond of 385 m^2) and operated as stagnant systems, except during water renewal. Rice bran was used to feed the fish at a rate of 2.5 g fish^{-1} day^{-1}, but with some feeding cuts to avoid overfeeding and water quality problems. The pond was dredged at the end of the experiment.

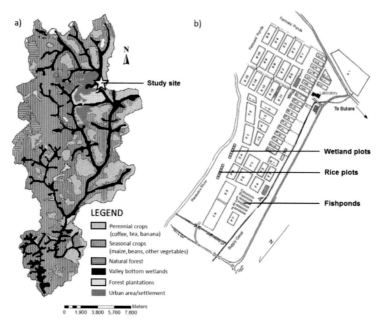

Figure 5-1. Map showing Rwasave fishpond station in Migina catchment (a) and Rwasave fishpond station layout with different experimental plots (b).

Rice and wetland plots were enclosed within 40 cm high peripheral clay dikes to prevent uncontrolled surface water flows. Water entered and left the experimental plots through pipes where water flow measurement was feasible. While rice farming involved practices like ploughing, weeding and fertilizer application, wetland plots were not subject to any treatment except water supply. Water, sediment, nitrogen and phosphorus flows were measured (Figure 5-2) and their corresponding activities (land preparation, stocking/planting, feed/fertilizer application, harvesting and dredging) in each LULC type were recorded. More details about the experimental setup are reported in Chapter 4 (Uwimana *et al.*, 2018b).

5.2.3 Data collection

5.2.3.1 Hydrology

Hydrological flows were measured from September 2012 to May 2013 and included inflows (surface inflow and rainfall), outflows (surface outflow and evaporation) and groundwater exchange. Results of these measurements were reported in Uwimana *et al.* (2018b). In short, fishponds used two times less water than rice farms and wetlands, probably as a result of the clay lining in the ponds that reduced water loss through seepage. This explains why seepage from fishponds was about three times less than in rice and wetland plots. Rainfall contributed 65% of the water flow into fishponds, but much less in rice (35%) and wetland plots (25%). Evaporation contributed more to water loss in fishponds (41%) than in rice (23%) and wetland plots (20%). With the exception of pond draining at the end of the farming season, and infrequent water renewal, there was no effluent from fishponds. Rice farms and natural wetlands lost most of their water supply (70-80%) through surface and subsurface outflow. The water flows were used to calculate loads of sediment and nutrient pathways (see next sections).

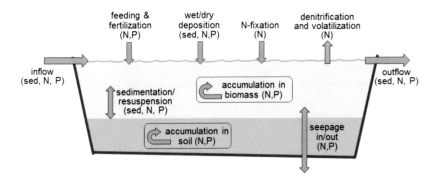

Figure 5-2. Conceptual model for sediment and nutrient pathways in fishponds, rice plots and wetland plots. Letters in parentheses indicate the units in which flows were quantified (s = sediment, N = nitrogen, P = phosphorus). For further explanation, see text.

5.2.3.2 Sediment and nutrient concentrations and loads

Sediment and nutrient loads were quantified according to conceptual diagram shown in Figure 5-2. This was done by combining the water flow estimations with determinations of sediment, nitrogen and phosphorus concentrations in the various flows. Concentrations and loads were determined as follows:

Inflow and outflow - Sediment concentration in inflow, water column and outflow was measured using a Portable colorimeter DR/890 (HACH, Colorado, USA) and sediment loads (g m^{-2} d^{-1}) calculated as a product of the total suspended solids (TSS) concentration (mg L^{-1}) and the water discharge (mm d^{-1}) over the plot surface area. Inflow and outflow water samples were collected and preserved by acidification to pH = 1-2, using sulfuric acid (0.1 N). Total nitrogen (TN) and total phosphorus (TP) concentrations were measured using the persulfate digestion method (APHA 2005). The nutrient flow rates (g m^{-2} d^{-1}) were calculated as a product of daily water discharge and the corresponding TN and TP concentrations per plot surface area (m^2). The resulting seasonal flow rates were summed up on a monthly basis (g m^{-2} mo^{-1}).

Feeding and fertilization - The amount of fish feed (rice bran) and fertilizer (NPK and urea) applied to ponds and rice plots were continuously recorded during the experimental period. Feeding or fertilizer application rates (g m^{-2} d^{-1}) were calculated as the total weight of the applied feed or fertilizer over the farming period divided by the number of days and the pond or plot surface area (m^2). Feed and fertilizer samples were dried, ground and analysed for TN and TP using the Kjeldahl digestion method (ICARDA 2001, APHA 2005). TN and TP feeding or fertilization rates (g m^{-2} d^{-1}) were calculated as a product of feeding or fertilization rate (g m^{-2} d^{-1}) and TN or TP content in feed or fertilizer.

Atmospheric deposition - Samples for atmospheric deposition were collected every 7 days using a trap supported at 1 m above the ground containing 1 litre of distilled water (Figure 5-3c). Trapped samples were analysed for sediment content (g) using a portable colorimeter DR/890 (HACH, Colorado, USA) and TN and TP content (g) using the persulfate digestion method (APHA 2005). Sediment and nutrient deposition rates (g m^{-2} d^{-1}) were calculated from trap sediment and TN or TP content, divided by the trap surface area (m^2) and collection period (d).

Figure 5-3. Data collection methods for sedimentation rate in fishpond (a), sediment drying in trap (b), atmospheric disposition sampling (c), sampling of sedimentation rate in rice farm and a piezometer around (d) and installation of a lysimeter in experimental plot (e) for percolate collection.

Sedimentation - Sedimentation rates were calculated as the quantity of sediments settled in the sediment traps over the settling period. Sediment traps in the fishponds consisted of cylindrical plastic containers attached to a cement base with strings attached to the outer edge (Figure 5-3a, b). The cement base helped in lowering and stabilizing the trap on the pond bottom, while the strings helped in lifting the trap. Cone-shaped plastic objects were used as sediment traps in the rice farms and wetland plots (Figure 5-3d). They were immobilized into the soil using the cone vertex and strings were attached to the outer edge to facilitate lifting of the trap. A simple walkway was constructed inside each plot to facilitate placement of the traps and sampling (Figure 5-3a). To obtain representative information of sedimentation rate along the fishpond (55 m long), three sediment traps were used in each pond: one 5 m from the inlet, the second in the middle, the third 5m from the outlet. Because rice and wetland plots were small (10 m wide and long) we used one sediment trap in the middle of each rice farm and wetland plot. The traps were left in the plot for around one month and collected sediments were dried in an oven at 65°C for 48 hours and then weighed. Sedimentation rates (g m^{-2} d^{-1}) were calculated as the weight of dried sediments over the trap surface area and trapping period.

Seepage in/out - Seepage rates were calculated using the water mass balance and were reported in Uwimana *et al.* (2018b). Percolates (groundwater) were collected from the underground drains connected to the lysimeter and piezometers that were installed at a depth of 50 cm below the ground in different plots (Figure 5-3e).

Gaseous exchanges of nitrogen - Ammonia volatilization in the fishponds, rice and wetland plots was estimated based on the aqueous ammonia, pH and temperature, measured hourly for three consecutive days. The calculation was based on the equilibrium between unionized free ammonia (NH_3), pH and temperature (Emerson *et al.* 1975) that relates the amount of free ammonia (potential for volatilization) to pH and temperature.

Denitrification was estimated based on data collected by Murekatete (2013) in the same experimental plots, using the acetylene inhibition method (Sørensen, 1978). This method inhibits the production of dinitrogen (N_2), transforms nitrates (NO_3) into dinitrogen oxide (N_2O) that is measurable using gas chromatography. This method is simple and less expensive and good for potential denitrification, hence it gives the maximum rates of denitrification compared to natural denitrification rates (Klemedtsson, 1990; Knowles, 1990).

Nitrogen fixation can be measured by N balance studies as the difference between easily measurable outputs and input during the experimental period. It can also be measured by acetylene-reducing activity (ARA) or by determining the N derived from the air. In this study we used literature data on nitrogen fixation in rice fields (Roger and Ladha, 1992).

Accumulation in biomass - TN and TP uptake by fish, rice and grass were assessed based on biomass growth. Fish, rice plant and grass samples were taken at the start of the experiment, and then periodically until harvest. Forty fish from each pond were weighed (g) and measured (cm) once per month. For rice and wetland plots, rice and grass belowground and aboveground biomass were taken in three quadrats of 1 m^2 in each experimental plot at the start, and then monthly until the end of the experiment. All fish and plant samples were dried in an oven at 65 °C for 48 hours and weighed. The biomass growth (g dry matter m^{-2} d^{-1}) was calculated from start to end of the experiment (259 days for fish; and 170 and 191 days for rice; and 274 days for wetlands). Dried and ground samples were analysed in terms of TN and TP content using the Kjeldahl digestion method (ICARDA 2001; APHA, 2005). TN and TP content were calculated as the product of the weight of fish, rice and wetland biomass and their respective concentrations. TN and TP uptake rate (g m^{-2} d^{-1}) was calculated as the difference between the TN or TP quantity at the end and at the start of the experiment over the surface area (m^2) and number days (d) of biomass growth.

Change in soil storage - The soil TN and TP content were measured at the start and end of the experiment. Zigzag soil sampling method (ICARDA, 2001) was used to collect soil samples at different points of each plot in the soil horizon of 30 cm depth. Samples were mixed to get a representative sample that was analysed for TN and TP concentration (%). Soil nutrient content was calculated based on soil concentration (%), bulk density and thickness of the soil layer. Change in soil nutrient storage was calculated as the difference between the soil content at the start and end of the experiment, as follows:

$$NC_A = [N, P] * BD * T$$

where NC$_A$ = soil nutrient content per surface area (g m^{-2}), $[N, P]$ = soil nutrient concentration (%), BD = bulk density (Mg m^{-3}), T= thickness of the soil layer (m).

Sediment and nutrient retention - Because we did not measure gaseous N exchanges, we did not include these in the monthly estimates for N retention. However, we did compare the resulting N retention to the gaseous exchange estimates afterwards. Nutrient accumulation in biomass was also not included in the retention estimates but compared to nutrient removal in rice and fish harvest at the end of the culture period. Apparent sediment and nutrient retention were thus calculated for every month, as follows:

- Sediment retention (g dm m^{-2} d^{-1}) =
  inflow + atmospheric deposition + feed input - outflow

- N retention (g N m^{-2} d^{-1}) =
  inflow + atmospheric deposition + feed/fertilizer input - outflow - seepage

- P retention (g P m^{-2} d^{-1}) =
  inflow + atmospheric deposition + feed/fertilizer input - outflow - seepage

5.2.3.3 Data analysis

To compare sediment and nutrient flows among the three LULC systems, the monthly load estimates were analysed using an approach for analysis of longitudinal data presented by Everitt and Hothorn (2010). Linear mixed effects models were formulated, using land use type and month as fixed variables and individual plots as a random effect to account for the dependence of the repeated measurements. From random intercept and random slope models, the model with the lowest AIC was selected. Mean sediment, N and P loads in the land use types for the whole period were compared using least square means with Tukey adjustment. All analyses were done with log10-transformed values as residual analysis showed more normal distribution of the residuals. All calculations were done using the R software version 3.3.1 (R Core Team, 2016), and in particular function lme for estimating the mixed models, and function lsmeans for posthoc analysis.

5.3. Results

5.3.1. Nutrient analysis

Table 5-1 shows the concentrations (mg L^{-1} or % on dry matter basis) of N and P as measured in different flows of the fishpond, rice farm and wetland systems. Mean N and P inflow concentrations were higher in rice farms (5.6 mg L^{-1} of N and 0.16 mg L^{-1} of P) than in fishponds (2.3 mg L^{-1} N; 0.11 mg L^{-1} P) and wetlands (1.6 mg L^{-1} N; 0.12 mg L^{-1} P). Differences in N concentrations in the outflow followed the same trend with higher N concentration in rice farm effluent (5.1 mg L^{-1}) than in fishponds (4.5 mg L^{-1}) and wetlands (1.1 mg L^{-1}). P was higher in the fishpond effluent (0.32 mg L^{-1}) than for the rice farms (0.24 mg L^{-1}) and wetland (0.08 mg L^{-1}). In percolate, N was higher in the wetland (3.3 mg L^{-1}) than in the fishponds (3.0 mg L^{-1}) and rice plots (1.9 mg L^{-1}), while P was higher in fishpond percolate (0.40 mg L^{-1}) than in rice (0.31 mg L^{-1}) and wetlands (0.27 mg L^{-1}). In terms of fertilization, higher concentrations were observed in fertilizer applied to rice farm (30.5% N and 17.0% P) than in feeds to fishponds (4.4% N and 3.4% P). Nutrients in sedimentation were higher in wetland plots (4.2% N and 3.0% P) than in fishponds (3.4% N and 2.2% P) and rice plots (3.4% N and 0.9% P). Nutrients in soil were higher in fishponds (3.0% N and 1.2% P) than in wetlands (2.2% N and 0.37% P) and rice plots (1.3% N and 0.28% P). In biomass, nutrient concentration was higher in wetland plant dry matter (5.3%) than in fish (4.7%) and rice (4.5%). P concentration was higher in fish (2.8%) than in wetland (1.8%) and rice (1.7%) vegetation.

5.3.2. Seasonal variation in sediment and nutrient loads

5.3.2.1 Sediment

There was a strong seasonal variation of sediment loads in the different LULC systems (Figure 5-4). In fishponds, high sediment loads in the inflow (16 g m^{-2} d^{-1}) were observed in September during filling of the ponds. High sediment loads in the outflow were observed in February (10 g m^{-2} d^{-1}) during water renewal, and in May 2013 (2251 g m^{-2} d^{-1}) during pond dredging. Sedimentation rates in fishponds varied in the range of 15-43 g m^{-2} d^{-1}, with highest values (36-43 g m^{-2} d^{-1}) in the period December 2012-February 2013 during plankton blooms. Atmospheric deposition of sediment was highest (0.5-0.6 g m^{-2} d^{-1}) in dry periods (end of the dry season in September and in February-March), and decreased in the rainy periods (December, April-May).

Table 5-1. Mean N and P concentrations (mg L^{-1} in water) and (% in dry matter) in sediment, biomass and soil in different land use/land cover (LULC) systems (rice plots, fishpond and wetland plots) in Rwasave fish farm, Rwanda in the period September 2012 - May 2013. sd = standard deviation, N = no. of samples. For further explanation, see text.

Process/flow	Material (unit)	LULC system	Nitrogen			Phosphorus		
			Mean	sd	N	mean	sd	N
Inflow	Water (mg L^{-1})	Rice	5.61	3.88	34	0.16	0.10	34
		Fish	2.31	1.79	42	0.11	0.11	42
		Wetland	1.57	1.55	42	0.12	0.15	42
Outflow	Water (mg L^{-1})	Rice	5.10	5.04	34	0.24	0.17	34
		Fish	4.51	5.87	42	0.32	0.39	42
		Wetland	1.14	0.96	42	0.08	0.09	42
Seepage	Water (mg L^{-1})	Rice	1.89	1.51	7	0.31	0.12	7
		Fish	2.97	1.16	9	0.40	0.27	9
		Wetland	3.34	1.13	7	0.27	0.35	7
Feeding	Feed (% dm)	Fish	4.43	0	5	3.37	0	5
Fertilization	Fertilizer (% dm)	Rice	30.53	15	4	17	0	2
		Wetland	-	-	-	-	-	-
Sedimentation	Sediment (% dm)	Rice	3.39	0.96	34	0.89	1.61	34
		Fish	3.40	0.47	42	2.18	1.16	42
		Wetland	4.16	1.05	42	3.03	3.15	42
Accumulation in soil	Soil (% dm)	Rice	1.29	0.24	25	0.28	0.03	25
		Fish	2.96	0.26	10	1.19	0.15	10
		Wetland	2.16	0.33	12	0.37	0.03	12
Accumulation in biomass	Biomass (% dm)	Rice	4.49	1.03	7	1.72	0.60	7
		Fish	4.66	0.006	9	2.83	0.22	9
		Wetland	5.34	0.67	7	1.83	0.81	7

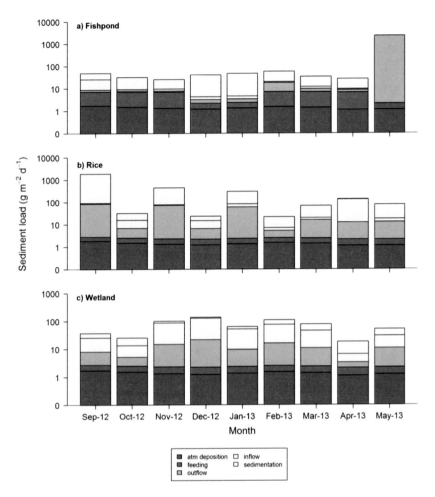

Figure 5-4. Seasonal variation of sediment loads in different land use/land cover systems (a: fishponds, b: rice plots, and c: wetland plots) in Rwasave fish farm, Rwanda in the period September 2012 - May 2013. For further explanation, see text.

In the rice plots, farming activities (ploughing, levelling and seedling transplanting in August-September 2012 and January 2013, as well as weeding in November 2012 and March 2013) loosened and resuspended soil and sediment particles leading to much higher sedimentation rates (49-1771 g m^{-2} d^{-1}) and outflows (13-82 g m^{-2} d^{-1}) compared with fishponds. The end of rice farming was characterized by low outflows (3.5-7.4 g m^{-2} d^{-1}), due to the absence of human disturbing activities and the dense rice crop that trapped sediment particulates.

In wetland plots, high sediment loads in the inflows (34-105 g m^{-2} d^{-1}) were observed in the period of November 2012-March 2013, following the cleaning of the water supply canals. Both the outflows (6.5-19 g m^{-2} d^{-1}) and the sedimentation rates (32-35 g m^{-2} d^{-1}) showed the same trend with higher loads during that period.

5.3.2.2 Nitrogen

There was a strong seasonal variation of N loads among different LULC types (Figure 5-5). In fishponds inflows, TN load was highest in September during water filling (318 mg N m^{-2} d^{-1}). Water renewal was associated with very low TN loads (<5 mg m^{-2} d^{-1}). TN discharge in fishpond outflows was very low to inexistent throughout the farming season (<1.4 mg m^{-2} d^{-1}), except during lowering of the water (fish control, February) and draining/dredging of the ponds (in May 2013). TN sedimentation dominated the flows during the farming season, with peaks in December 2012-February 2013 (1486-1693 mg m^{-2} d^{-1}) when plankton bloomed and feed application had to be stopped (24.4 mg m^{-2} d^{-1}). Feed application was resumed at the end of February after water renewal. Atmospheric deposition varied too with seasons, with higher values (174 mg m^{-2} d^{-1}) at the end of the long dry season in September and lower values (5.4 mg m^{-2} d^{-1}) at the end of the rainy season. Nitrogen uptake by fish seasonally varied in a narrow range (5.3-12 mg m^{-2} d^{-1}), with higher values in the early stages of fish farming.

High seepage (3.0 mg m^{-2} d^{-1}) was observed in November 2012, and January and April 2013 following high water levels in the ponds after rainfall events. Negative values (-2.7 to -11 mg m^{-2} d^{-1}) of seepage (groundwater flow to fishpond) were observed in February and May 2013 when water level in the ponds was low, following periods of water lowering and dredging. Ammonia volatilization, denitrification and nitrogen fixation in fishponds were estimated at 5.5, 5.4 and 4.7 mg N m^{-2} d^{-1}, respectively.

In rice plots, high TN loads in inflows were observed in November 2012 (102 mg m^{-2} d^{-1}) and April 2013 (194 mg m^{-2} d^{-1}), following high storm water flows from the catchment. This period was also associated with high TN load in the outflows (51-73 mg m^{-2} d^{-1}). N fertilizer application was high in September 2012 and in January 2013 during seedling transplanting (425 mg m^{-2} d^{-1}) and in November and March after weeding (230 mg m^{-2} d^{-1}). Following the disturbance of the soil during ploughing and weeding, a large amount of TN was observed in sediment traps right after ploughing and weeding (1068-64392 mg N m^{-2} d^{-1}). Ammonia volatilization, denitrification and nitrogen fixation in the rice plots were estimated at 3.0, 27.7 and 2.2 mg N m^{-2} d^{-1}, respectively.

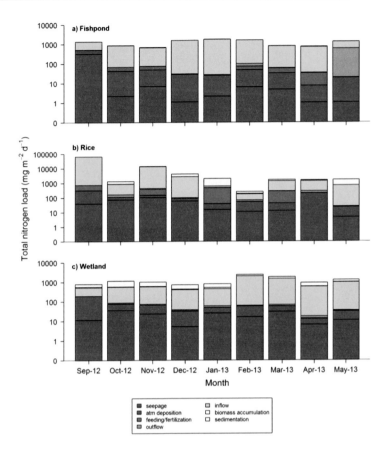

Figure 5-5: Seasonal variation of nitrogen load in different land use/land cover systems (a: fishponds, b: rice plots, and c: wetland plots) in Rwasave fish farm, Rwanda in the period September 2012 - May 2013. For further explanation, see text.

In wetland plots, there was no marked seasonal variation of different N flows (inflow, outflow, settling, seepage and biomass), probably because of the absence of human activities in the wetland. However, higher TN loads were collected in the sediment traps in February 2012 (1969 mg m^{-2} d^{-1}) and March 2012 (1413 mg m^{-2} d^{-1}), following considerable input from falling leaves that decomposed and generated high N in the outflow (10.3 mg m^{-2} d^{-1}) in March 2012. Compared with rice plots and fishponds, the wetland plots had higher N seepage loads (74.5 mg m^{-2} d^{-1}). Ammonia volatilization, denitrification and nitrogen fixation in the wetland plots were estimated at 1.4, 75.8 and 16.2 mg N m^{-2} d^{-1}, respectively.

5.3.2.3. Phosphorus

As for TN, there was a strong seasonal variation of TP loads in the different LULC types (Figure 5-6). High TP loads were observed in fishpond inflows (13 mg m^{-2} d^{-1}) in September during water filling, and very low loads (0.4 mg m^{-2} d^{-1}) during water renewal. TP loads in fishpond outflows were related strongly to pond management activities: they were inexistent when the pond operated as a stagnant system without outflow, very low (<1.4 mg m^{-2} d^{-1}) during water discharge (water renewal and fish control), and very high (50 mg m^{-2} d^{-1}) during pond draining at the end of the farming period (May 2013). Sedimentation dominated all other TP flows during the farming season (327-1420 mg m^{-2} d^{-1}), with peaks during the plankton blooms in January-February 2013 (729-1420 mg m^{-2} d^{-1}). As for TN, atmospheric deposition of TP varied with seasons, higher in dry periods (4.5 - 7.7 mg m^{-2} d^{-1}) and lower (1.6 mg m^{-2} d^{-1}) at the end of rainy season. As for TN, TP uptake in biomass (3.2-6.9 mg m^{-2} d^{-1}) was highest in the early stages of fish farming, and highest seepage was observed in November 2012 (1.8 mg m^{-2} d^{-1}) and January 2013 (1.9 mg m^{-2} d^{-1}) with negative seepage in February 2012 and May 2013 (-0.7 and -2.6 mg m^{-2} d^{-1}, respectively).

Similar to TN, high TP loads in inflows of rice plots were observed in November 2012 (2 mg m^{-2} d^{-1}) and April 2013 (46 mg m^{-2} d^{-1}), following high storm water flows from the catchment. This period was also associated with high TP load in the outflows (1.9 mg m^{-2} d^{-1} in November 2012, and 10 mg m^{-2} d^{-1} in April 2013). Phosphorus applied in fertilizer was high in September 2012 and January 2013 during seedling transplanting (425 mg m^{-2} d^{-1}). A large amount of TP was observed in sediment traps right after ploughing and weeding (6780 mg m^{-2} d^{-1} in August 2012 during ploughing and 1368 mg m^{-2} d^{-1} in October 2012 during weeding).

In wetland plots, as for TN, there was no remarkable seasonal variation of TP flows (inflow, outflow, settling, seepage and biomass). Higher TP loads were collected in the sediment traps in February and March 2012 (624 and 1299 mg m^{-2} d^{-1}) following considerable input from falling and decomposing leaves. In contrast to TN seepage that was highest in wetland plots, TP seepage was higher in rice plots (2.5 mg m^{-2} d^{-1}) than in wetlands (1.8 mg m^{-2} d^{-1}) and fishponds (0.7 mg m^{-2} d^{-1}).

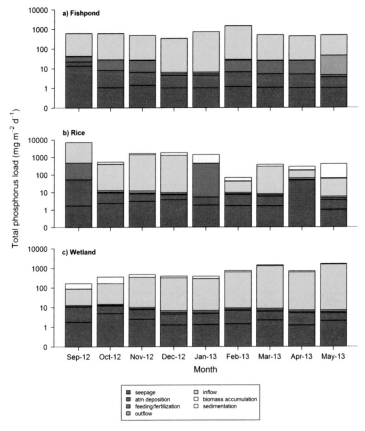

Figure 5-6: Seasonal variation of phosphorus load in different land use/land cover systems (a: fishponds, b: rice plots, and c: wetland plots) in Rwasave fish farm, Rwanda in the period September 2012 - May 2013. For further explanation, see text.

5.3.3 Differences among LULC types

Table 5-2 shows the mean loads of sediment, TN and TP over the whole experimental period in the various flows of the different LULC systems. In inflow, significantly higher sediment loads were observed in wetlands (43 mg m^{-2} d^{-1}) due to very turbid water from the upstream canals cleaning activities. Relatively low loads were observed in inflows to rice plots (13 mg m^{-2} d^{-1}) and fishponds (2.2 mg m^{-2} d^{-1}). Over the whole period, sediment loads were higher in effluent from fishponds (256 g m^{-2} d^{-1}) than in effluent from rice plots (34.7 g m^{-2} d^{-1}) and wetlands (8.8 g m^{-2} d^{-1}). However, when excluding the draining of fishponds, rice plots had the highest sediment outflow (see Table 5-2). Feed contributed 3.1 g m^{-2} d^{-1} of suspended solids in fishpond. Sedimentation was higher in rice plots (414 g m^{-2} d^{-1}) than in fishponds (25.5 g m^{-2} d^{-1}) and wetlands (18.2 g m^{-2} d^{-1}). Atmospheric deposition was 0.39 g m^{-2} d^{-1} across the different LULC types.

87

Table 5-2. Mean (± se) sediment loads (g dry matter m^{-2} d^{-1}) in different land use/land cover LULC systems (rice plots, fishponds and wetland plots) in Rwasave fish farm, Rwanda in the period September 2012 - May 2013. se = standard error, N = no. of samples. Within flow types, LULC systems sharing the same superscript are not significantly different from each other (linear mixed effects model, and post-hoc pairwise comparison test, P< 0.05). For further explanation, see text.

Process/flow	Material	LULC system	Sedimentation rate (g d^{-1} m^{-2})		
			Mean	se	N
Inflow	Water	Fish	2.2	± 0.70[a]	53
		Rice	13.1	± 6.0[b]	33
		Wetland	42.6	± 8.4[c]	46
Feed	Feed	Fish	3.1	± 0.38	53
	Fertilizer	Rice	-	-	-
		Wetland	-	-	-
Atmospheric deposition	Water	All	0.39	± 0.016	132
Outflow[1]	Water	Fish	256	± 156[a]	53
		Rice	34.6	± 7.0[b]	33
		Wetland	8.8	± 1.4[ab]	46
Sedimentation	Sediment	Fish	25.4	± 2.1[a]	53
		Rice	414	± 125[b]	33
		Wetland	18.2	± 3.8[a]	46

[1]Mean values for outflow without the last sampling in May (with pond drainage) were 1.7 ± 0.55[a] (Fish), 36 ± 7.4[c] (Rice), and 8.9 ± 1.5[b] (Wetland).

Nitrogen load in various flows was significantly different among the LULC systems (Table 5-3). N load through surface inflow was higher in the rice plots (57 mg m^{-2} d^{-1}) than in fishponds (40 mg m^{-2} d^{-1}) and wetland plots (19 mg d^{-1} m^{-2}). However, higher N load was observed in the outflow of fishponds (71 mg m^{-2} d^{-1}), compared to the N load in rice (23.9 mg m^{-2} d^{-1}) and wetland outflows (5.5 mg m^{-2} d^{-1}). Nitrogen seepage was higher in wetland plots (28 mg d^{-1} m^{-2}) than in rice plots (11.6 mg m^{-2} d^{-1}) and fishponds (8.6 mg m^{-2} d^{-1}). Nitrogen fertilization was higher in rice (147 mg m^{-2} d^{-1}) than in fishponds (17.4 mg m^{-2} d^{-1}) and wetlands (no fertilization). As a result of ploughing and weeding, activities that loosen and resuspend the soil and associated nutrients, N sedimentation was very high in rice plots (14254 mg m^{-2} d^{-1}) compared with fishponds (967 mg m^{-2} d^{-1}) and wetland plots (873 mg m^{-2} d^{-1}). Nitrogen accumulation in the soil was higher in fishponds (1.02 mg m^{-2} d^{-1}) than in wetlands (0.75mg m^{-2} d^{-1}) and rice plots (0.45 mg m^{-2} d^{-1}). Nitrogen uptake in biomass was higher in rice (662 mg m^{-2} d^{-1}) than in wetland grass (359 mg m^{-2} d^{-1}) and fish (7.4 mg m^{-2} d^{-1}). Ammonia volatilization was higher in fishponds (364 mg m^{-2} d^{-1}) than in rice (3 mg m^{-2} d^{-1}) and wetlands (1.4 mg m^{-2} d^{-1}).

Table 5-3. Mean (± se) nitrogen and phosphorus (mg N or P m^{-2} d^{-1}) in different land use/land cover LULC systems (rice plots, fishponds and wetland plots) in Rwasave fish farm, Rwanda in the period September 2012 - May 2013. se = standard error, N = no. of samples. Within flow types, LULC systems sharing the same superscript are not significantly different from each other (linear mixed effects model, and post-hoc pairwise comparison test, P< 0.05). For further explanation, see text.

Rate	Material	LULC type[3]	Nitrogen (mg N m^{-2} d^{-1})			Phosphorus (mg P m^{-2} d^{-1})		
			mean	Se	N	mean	se	N
Inflow	Water	F	40.0	± 19[a]	42	1.6	± 0.73[a]	42
		R	56.8	± 12[c]	34	3.9	± 2.5[a]	34
		W	19.2	± 3.1[b]	42	1.3	± 0.24[a]	42
Feeding or Fertilization[2]	Feed	F	17.4	± 1.7	42	13.2	± 1.3	42
	Fertilizer	R	147	± 31	34	100.0	± 31.4	34
		W	-	-	-	-	-	-
Atmospheric deposition[2]	Material from trap	All	43.4	± 16.7	9	3.9	± 0.67	9
Outflow[1]	Water	F	71.0	± 30[a]	42	4.9	± 2.1[a]	42
		R	23.9	± 5.9[b]	34	1.4	± 0.53[a]	34
		W	5.5	± 0.70[ab]	42	0.38	± 0.075[a]	42
Seepage	Water	F	8.6	± 1.9[a]	42	0.33	± 0.21[a]	42
		R	1.6	± 2.4[a]	34	2.4	± 0.42[b]	34
		W	28.0	± 3.1[b]	42	1.7	± 0.28[b]	42
Sedimen-tation	Material from trap	F	967	± 100[ab]	42	621	± 112[a]	42
		R	14254	± 4522[b]	34	1728	± 517[a]	34
		W	873	± 238[a]	42	636	± 159[a]	42
Accumulation in soil	Soil	F	1.02	± 0.016	24	0.41	± 0.016	24
		R	0.45	± 0.028	10	0.09	± 0.0018	10
		W	0.75	± 0.035	12	0.13	± 0.0029	12
Accumula-tion in biomass	Fish	F	7.4	± 0.40[a]	42	4.4	± 0.20[a]	42
	Rice	R	662	± 88.5[b]	34	270	± 43.5[c]	34
	Vegeta-tion	W	359	± 19.6[b]	42	121	± 9.2[b]	42

[1]Mean values for outflow without the last sampling in May (with pond drainage) were, for N: 3.9 ± 1.5[a] (Fish), 27 ± 6.5[c] (Rice), and 6.1 ± 0.77[b] (Wetland); for P: 0.22 ± 0.089[a] (Fish), 1.6 ± 0.59[b] (Rice), 0.40 ± 0.087[a] (Wetland); [2]No statistical comparison was done; F = fish, R= rice, W = wetland

In general, P loads followed a similar pattern as N loads (Table 5-3). As for N, P load was higher in rice farm inflow (3.9 mg m^{-2} d^{-1}) than in the inflow of fishpond (1.6 mg m^{-2} d^{-1}) and wetland (1.3 mg m^{-2} d^{-1}). However, P load was higher in the outflow of fishponds (4.9 mg m^{-2} d^{-1}) than in rice (1.4 mg m^{-2} d^{-1}) and wetland effluent (0.38 mg m^{-2} d^{-1}). In contrast with N load that was higher in percolate of the wetlands, P percolate was higher in rice plots (2.4 mg m^{-2} d^{-1}) than in wetlands (1.6 mg m^{-2} d^{-1}) and fishponds (0.33 mg m^{-2} d^{-1}). P fertilization load was higher in rice (100 mg m^{-2} d^{-1}) than in fishponds (13.2 ± 8.5 mg m^{-2} d^{-1}) and wetlands (no fertilization). Following soil and sediment disturbances from ploughing and weeding, P settling was higher in rice (1,729 mg m^{-2} d^{-1}) than in wetland (636 mg m^{-2} d^{-1}) and fishponds (621 mg m^{-2} d^{-1}). P accumulation in soil was higher in fishponds (0.41 mg m^{-2} d^{-1}) than in wetlands (0 mg m^{-2} d^{-1}) and rice plots (0.09 mg m^{-2} d^{-1}). Nitrogen accumulation in biomass was highest in rice (270 mg m^{-2} d^{-1}), compared with wetland grass (121 mg m^{-2} d^{-1}) and fish (4.4 mg m^{-2} d^{-1}).

5.3.4 Retention of sediment, nitrogen and phosphorus

The retention of sediment varied seasonally with LULC systems (Figure 5-7; Table 5-4). Seasonally, the fishpond sediment retention varied in the wide range of -7890 to 26 g m^{-2} d^{-1}, with positive values observed in September 2012 (17.8-25.7 g m^{-2} d^{-1}), during filling of the ponds and negative values (exports) observed in May 2013 (-664 to -7890 g m^{-2} d^{-1}) during drainage of the ponds. The lowering of the water during fish control in February 2013 was also associated with small exports (-2.1 to -11.4 g m^{-2} d^{-1}). Sediment retention in rice plots varied seasonally in the range -108 to 184 g m^{-2} d^{-1} with negative values (exports) observed in September, November and January during ploughing and weeding (<-108 g m^{-2} d^{-1}), and positive values observed in other periods. In wetlands, sediment retention was always positive, with the exception of two observations, one in December 2012 (-9 g m^{-2} d^{-1}) another in February 2013 (-6.6 g m^{-2} d^{-1}). On average, sediment retention was negative (export) for fishponds (-251 g m^{-2} d^{-1}) and rice plots (-21.2 g m^{-2} d^{-1}), but positive for wetlands (34.2 g m^{-2} d^{-1}). Because of the wide variation, these differences were not significant (P>0.05), but leaving out the last month with the pond drainage, wetlands had a significantly higher sediment retention than rice plots and fishponds (Table 5-4).

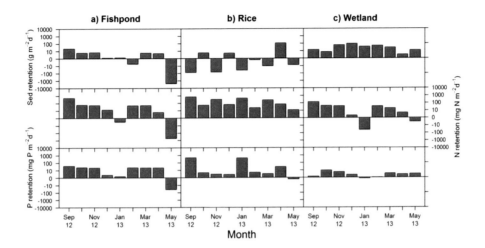

Figure 5-7. Mean apparent retention of sediment, nitrogen and phosphorus in three land use/land cover systems (a: fishponds, b: rice plots, and c: wetland plots) in Rwasave fish farm, Rwanda in the period September 2012 - May 2013. For further explanation, see text.

The retention of TN varied seasonally with LULC systems. TN retention was positive (11.7-912 mg m^{-2} d^{-1}) in the first three months of fish farming. Negative values (exports) were observed after three months of farming during fishpond drainage, with peaks in May at the end of farming (-351 to -857 mg m^{-2} d^{-1}). Due to high fertilization rate in rice plots and higher rice nitrogen uptake, the retention of TN was positive most of the time: only one case of export (-2.1 mg m^{-2} d^{-1}) was observed in April 2013 during heavy rainfall. In wetland plots, the retention of TN was predominantly positive (only five cases of TN export were observed in different periods). On average, the retention of TN was positive for all LULC (241 mg m^{-2} d^{-1} for rice plots, 37.0 mg m^{-2} d^{-1} for wetlands and 22.5 mg m^{-2} d^{-1} for fishponds). Taking biomass harvesting into account, TN retention in fishponds decreased slightly to 17 mg m^{-2} d^{-1}, but in rice plots changed from retention to export (to -1054 mg m^{-2} d^{-1}). In wetland plots, if biomass would be harvested, retention would change to -246 mg m^{-2} d^{-1}. This indicates that rice and wetland biomass contain substantial amounts of N that play an important role in the TN retention capacity of these systems, in contrast to fishponds where fish biomass harvesting does not influence TN retention much. Because of the relatively low values, including gas exchanges (N fixation, ammonia volatilization and denitrification) in the retention calculations did not affect the net retention for all LULC systems much.

Table 5-4. Mean (± se) sediment (g dry matter m^{-2} d^{1}), nitrogen and phosphorus (g N or P m^{-2} d^{1}) retention in different land use/land cover LULC systems (rice plots, fishponds and wetland plots) in Rwasave fish farm, Rwanda in the period September 2012 - May 2013. se = standard error, N = no. of samples. Within retention types, LULC systems sharing the same superscript are not significantly different from each other (linear mixed effects model, and post-hoc pairwise comparison test, P< 0.05). For further explanation, see text.

LULC system	Sediment (g m^{-2} d^{-1})			N (mg m^{-2} d^{-1})			P (mg m^{-2} d^{-1})		
	mean	Se	N	mean	se	N	mean	se	N
(for whole period)									
Fish	-251	± 157[a]	53	22.5	± 42.8[a]	42	13.6	± 3.3[a]	42
Rice	-21.2	± 9.7[a]	33	241	± 44.7[b]	34	112	± 33.9[b]	34
Wetland	34.2	± 7.7[a]	46	37.0	± 7.7[a]	42	2.8	± 0.51[a]	42
(without May 2013)									
Fish	4.7	± 1.1[a]	47	98.5	± 30.0[a]	37	20.2	± 1.9[a]	37
Rice	-22.2	± 10.3[a]	31	271	± 48.1[b]	30	127	± 37.6b	30
Wetland	37.7	± 8.8[b]	40	43.5	± 8.4[a]	36	2.9	± 0.59[a]	36

TP exports from fishponds were only observed during fishpond drainage in May (-26.8 to -61.7 mg m^{-2} d^{-1}). In rice plots, TP retention was positive most of the time (61.7 mg m^{-2} d^{-1}), probably due to high fertilizer input (100 mg m^{-2} d^{-1}) and less output, and became negative when harvesting was taken into account (-158 mg m^{-2} d^{-1}). The net retention of TP was positive for all study LULC systems and significantly higher in rice plots (112 mg m^{-2} d^{-1}) than in fishponds (13.6 mg m^{-2} d^{-1}) and wetland plots (2.8 mg m^{-2} d^{-1}). Taking biomass harvesting into account, TP retention decreased to 9.1 mg m^{-2} d^{-1} in fishponds and to -158 mg m^{-2} d^{-1} and -118 mg m^{-2} d^{-1}) in rice and wetlands plots, respectively, indicating (as for N) that rice and wetland biomass contain substantial amount of phosphorus.

5.4 Discussion

In this study we assessed the effect of conversion of natural wetlands to rice and fish farming on sediment, N and P loads. We used mesocosm experimental wetland plots (as a model for natural land cover), rice plots (the most common and productive farming system) and fishponds (fish production with the potential of storing water and nutrients) to compare water and nutrient cycling in these LULC types. There were clear differences in flows among the three LULC types that were related to their characteristics and management activities.

Sediment, N, and P discharges (concentrations and loads) to the outflow were generally lower in the wetland plots than in the fishponds and rice plots, confirming that wetland can retain sediment and nutrients (Verhoeven *et al.* 2006, Reddy and Delaune 2008 and Kadlec and Wallace 2009, Uwimana *et al.* 2018b). On average, TSS retention was positive in wetlands. While wetland sediment retention can be explained by the dense vegetation land cover, low release of sediments can be attributed to fewer disturbances of the wetland by humans. However, wetland lost higher N and P loads to the groundwater by seepage (see Tables 1 and 3) as a result of the higher seepage flows from wetlands than from fishponds and rice plots (Uwimana *et al.* 2018b).

Management practices (ploughing and weeding in rice, water renewal and draining in fishponds) created strong seasonal variation in sediment flows in the different LULC types. While ploughing and weeding occur at the early stages (first three months) of rice farming, water renewal and dredging occur in the middle and end of fish farming. This makes the fishponds a temporal storage of sediments at the early stages of farming and a source of sediments towards the end, in contrast with rice farming that generates sediments early in the farming period (during land preparation) and traps sediment at the end of farming when vegetation cover is dense. Uwimana *et al.* (2018) also reported sediment retention by rice farms at the late stages of farming, while Nhan *et al.* (2008) and Kufel (2012) reported temporal retention of sediment by fishponds through accumulation in sediments, which are removed at the end of the farming season.

Nutrient loads varied across the different LULC types according to land use characteristics. The higher biomass in rice and wetlands compared to fishponds (2 and 89 times higher in rice than in wetlands and fishponds, respectively) reflected the higher nitrogen uptake in rice and wetland vegetation. Nitrogen recovery in rice can be around 40% (Schnier, 1995; Krupnik *et al.*, 2004) and in fish pond farming is only 5-6% (Nhan *et al.*, 2008). The higher ammonia volatilization in fishponds (260 and 121 times higher than in wetlands and rice farms, respectively) originates from the accumulation and degradation of nitrogenous feed waste and fish excreta whose emission peaks during the daytime hours when temperature, pH and air turbulence are high (Boyd and Tucker (1995, 1998; Avnimelech, 1998). Ammonia volatilization can also be an important nitrogen loss (up to 60%) from flooded rice farms where ammonia based fertilizer is spread over water (Fillery *et al.*, 1984).

Despite higher N input as fertilizer in rice farms (10 times higher than in fishponds) and the absence of fertilizer input in wetlands, the soil N storage in rice farms decreased at the end of the farming season, but increased in fishponds and wetlands. This probably reflects the higher nutrient uptake by the rice crop that can exhaust soil nutrients at low fertilizer application rates or inadequate soil conservation practices (Stoorvogel and Smaling, 1990; Henao and Baanante, 1999; Tittonell and Giller, 2013). In contrast to high nutrient uptake by rice, fish recover less nutrients from feed inputs as a part of the feed is not eaten by fish and accumulates in sediments instead (Kufel, 2012). Higher P discharges in fishpond outflows (3 and 12 times higher than in rice farms and wetlands, respectively) reflected the fishpond drainage (water renewal) and dredging at the end of farming season, similar to what has been reported from other studies in fishponds (Banas *et al.,* 2002; Konnerup *et al.,* 2011). The sub-surface P flow was higher in rice farms and wetlands (loose and permeable soil) than in the fishponds (compacted clayey soil). Despite the higher fertilization in rice (10 times higher than in the fishponds), soil storage in P decreased by 1.4% in the rice plots, but increased by 4.4% in the fishponds and by 1% in the wetland plots. The decrease in P soil storage in rice farms was attributed to higher biomass uptake (662 mg m^{-2} d^{-1}) that exhausts P in soil (Stoorvogel and Smaling, 1990; Henao and Baanante, 1999). The P soil storage increase in fishponds was attributed to low P uptake by fish (Kufel, 2012). P soil storage increase in wetland can be attributed to biomass senescence and decomposition and subsequent accumulation in wetland sediments.

Sediment resuspension was not measured in this study. From our field observations, resuspension seems to occur mostly in the early stages of rice farming during land preparation and transplanting. In fishponds, resuspension occurs at later farming stages following algal growth and disturbance of bottom sediment by fish movement and by shearing forces during water renewal and pond dredging. The suspended material settles back to the pond bottom, probably less than in rice plots because of the more mineral characteristics of suspended material in rice plots. In rice plots there is no need to remove the accumulated sediment, contrary to fishponds where dredging at the end of the farming period aims at the complete removal of sediments.

Assuming that there is no net change in suspended solids in the water column, sediment resuspension into the water column can be estimated from the difference between inputs (inflow, atmospheric deposition and feeding/fertilization) and outputs (outflow and sedimentation). The estimates showed a net negative resuspension in wetlands at the end of farming season (Figure 5-8). Negative resuspension suggests a net trapping of sediment (sink mechanism other than sedimentation), perhaps involving particulate attachment/adsorption or decomposition/dissolution of organic matter. In systems with a dense vegetation (such as in dense wetland vegetation or an established rice crop) sediments can be trapped with increased water residence time (Kadlec and Wallace, 2009).

In the rice plots, this process is disturbed during the early stages of rice farming (land preparation and transplanting), but is restored during the latter part of the farming period when the rice crop is dense. In fishponds, resuspension occurs at later farming stages following algal growth and bottom sediment disturbances during water renewal and pond dredging. Two cases of negative resuspension were observed in fishponds. These reflect the decomposition and dissolution of suspended material (algae and feeds). While 90 % of the suspended material from rice farming activities settled back, only 52% of suspended material settled back in fishponds. This may be due to the characteristics of the sediment which is more mineral in rice farms and more organic in fishponds. Also, in fishponds the management practices aim at removing most of the sediments at the end of farming season, contrary to rice farming which keeps the effluent to a minimum. This makes fish farming sediment outflow 7 and 28 times higher than sediments from rice farming and natural wetlands, respectively.

a) Fishpond

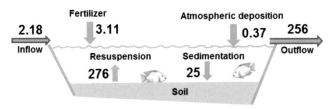

b) Rice

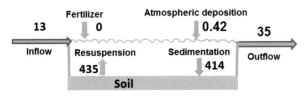

c) Wetland

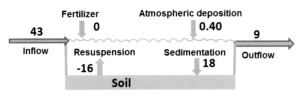

Figure 5-8: Suspended solid processes (g m⁻² d⁻¹) in different land use/land cover systems (a: fishponds, b: rice plots, and c: wetland plots) in Rwasave fish farm, Rwanda. Resuspension was calculated as the difference between gains (inflow, feed/fertilizer inputs, and atmospheric deposits) and losses (sedimentation, outflow). For further explanation, see text.

It was not possible to establish the mass balance for N and P due to unknown flows such as diffusion (for N and P) denitrification and N-fixation (for N). However, it is well documented that denitrification is higher in wetlands than in rice farm and fishponds due to the favourable environment of wetlands for active denitrifiers, notably the presence of organic carbon (McClain *et al.* 2003). Several studies (Ladha and Boonkerd, 1988; Roger *et al.*, 1988; Roger and Ladha, 1992) have also reported higher N fixation in wetlands than in rice farms and fishponds due to the presence of organic carbon and the absence of mineral nitrogenous fertilizers.

This study has shown that fishponds store water and accumulate sediment, N and P in the water column and sediment which are drained during water renewal, harvesting and dredging. It has also shown that N and P accumulate in the soil of fishponds and wetlands, as opposed to rice plots where N and P are taken up by the rice crop and exported through harvesting. This confirms that agricultural areas, wetland vegetation and ponds can alternately function as sinks and as sources for sediments and nutrients (Bolstad and Swank 1997; Hirsch 2012; Uwimana *et al.* 2017). Although this study was limited to a short period (August 2012-June 2013), data on the atmospheric sediment and nutrient deposition showed a seasonal pattern with higher loads in the dry periods and lower loads at the end of the rainy season. This further supports the concept of a build-up and washout mechanism (Uwimana *et al.* 2017) by which sediment, N and P accumulate in the atmosphere in form of particulates and gases, and are washed down through rainfall.

Within catchments, it is worth considering a combination of practices that can seasonally balance sink and source functions of different land uses for water, sediment, N and P. For example, water stored, and sediments and nutrients accumulated during fish farming can be drained at the end of the farming period to be used as a source of water, soil and nutrients for rice farming. The nutrient-rich pond bottom sediments can also be recycled within the pond system through rotational rice and fish farming in the ponds. Natural wetlands can be used to intercept sediments released by rice farming. Natural wetlands can rotationally be converted into rice farming and vice versa to balance nutrient accumulation and exhaustion of the soil. To reduce sediment and nutrient effluents from ploughing, leveling, transplantation and weeding of rice, it is important to keep water in-farm until the sedimentation is complete.

5.5 Conclusion

This study assessed the effect of conversion of natural wetlands to rice and fish farming on sediment, N and P loads. There were significant differences in loads related to LULC system characteristics, management practices, and seasons. In rice farming, resuspension and discharge of high sediment, N and P loads to the outflow were caused by ploughing and weeding in the early stages (first three months) of the culture period, while in fish farming water renewal and dredging during the middle and end of the fish farming period caused the discharge of nutrients and sediment. Nutrient uptake in fishponds was very low, compared with rice and wetland vegetation. This resulted in an accumulation of N and P in fishponds and wetland soil, and a decrease in storage (exhaustion) of N and P in rice soil. Water, sediment, N and P accumulate in the water column during a farming period and are discharged during water renewal, harvesting and dredging. To optimize the efficient use and recycling of water, sediments, N and P within a catchment it is important to consider a combination of land use types and practices: sediment and nutrient rich fish farming effluent can be recycled into rice farming, pond bottom sediments can be recycled through rotational rice farming, and natural wetlands and rice farms can interchangeably be converted into one another to balance the accumulation and exhaustion of the soil nutrients. Sediment and nutrient discharges from land preparation and transplanting of rice should be kept in-farm as much as possible until sediment settling is complete.

6. Effects of wetland conversion into farming on water quality, sediment and nutrient retention - synthesis, conclusions and recommendations

6.1 Synthesis of the findings

Despite the international recognition of the importance of wetlands and their ecosystem services, and international agreements for their protection (Ramsar Convention on Wetlands, 2018), contradictory policies and practices lead to the conversion of wetlands to farmland throughout eastern, western and southern Africa (Wood and Dixon, 2002; Adekola *et al.* 2012; Wood *et al.* 2013). This study investigated the impacts of conversion of natural valley-bottom wetlands in southern Rwanda to farming land on the retention of sediment, nitrogen (N) and phosphorus (P) in catchments and on the water quality of streams. The study used several approaches ranging from landscape-scale synoptic surveys at catchment, sub-catchment and river reach scale (Chapters 2 and 3), to controlled experiments at the scale of individual plots (Chapters 4 and 5). Data collection included land use and land cover (LULC: crop farms, fishponds and dams, forest and wetlands), landscape features (slope, area, length, and population density), rainfall, water discharge and water quality (N, P, suspended solids, dissolved oxygen, conductivity, pH and temperature).

The results from the landscape-scale synoptic surveys at catchment and reach scale (Chapters 2 and 3) showed that the temperature, pH, electrical conductivity (EC) and dissolved oxygen concentration (DO) decreased, and total suspended solids (TSS) increased with river discharge. N and P accumulated in the catchment during the dry season and washed into the water courses during the early stages of the higher flows, with subsequent lower concentrations at the end of the rains due to dilution. Human activities like cleaning of canals, ploughing, weeding, fertilizer application, fishpond drainage and dredging were the main source of sediments and nutrients in water when these activities coincided with high flow. Within a period of one year, large proportions of the annual loads of TSS, TP and TN (93%, 60% and 67%, respectively) were transported during 115 days with rain. Generally, rice and vegetable farming were a net source of nutrients and sediments, whereas grass/forest and ponds/reservoir generally were a net sink of nutrients and sediments. Seasonally, each of the studied reaches (rice farms, vegetable farms, ponds/reservoirs and grass/forest) shifted from being a source to a sink for nutrients and sediments, although overall reaches dominated by rice and vegetable farming contributed more to the export than reaches dominated by ponds/reservoir and grass/forest.

The small-scale experimental studies (Chapters 4 and 5) showed that water renewal and dredging in the middle and end period of fish farming generate and discharge large amounts of sediment, N and P in the outflow of these systems, in contrast to rice farming that generated sediments and nutrients early in the farming period and trapped them at the end of the farming season. Despite fertilizer input into rice farms (10 times higher than in fishponds) and the absence of fertilizer input into wetlands, nutrient soil storage decreased in rice farms (by 4.7 and 1.4% for N and P, respectively), but increased in fishponds (by 3.3% and 4.4% of N and P, respectively) and in wetlands (by 3.8% and 1% of N and P, respectively). The decrease in nutrient soil storage was attributed to higher N and P uptake in rice plots (on average 662 and 270 mg $m^{-2}d^{-1}$ of N and P, respectively) than in wetlands (359 and 121 mg $m^{-2} d^{-1}$ of N and P) and fishponds (7.4 and 4.4 mg $m^{-2} d^{-1}$ of N and P).The increase in nutrient soil storage in fishponds was attributed to lower uptake into fish biomass, with most of the nutrients accumulating in the pond sediment. The increase in nutrient soil storage in wetlands was, in the absence of biomass harvesting, attributed to biomass die-off and accumulation in the sediments.

To improve people's livelihoods and economic development while maintaining the water quality downstream of the farming area, the efficient recycling of water, sediments and nutrients in fishpond sediments should be considered. This could be achieved by recycling sediment and nutrient rich fish farming effluent as fertilizer in rice farming, and by rotational rice farming in pond bottom sediments. Natural wetlands and rice farms can interchangeably be converted into one another to balance the accumulation and exhaustion of the soil nutrients. Sediment and nutrient effluents generated during ploughing, levelling, and transplantation and weeding of rice should be kept on-farm until sediment settling is complete.

6.2 Scale issues

In sub-Saharan Africa, many river basin scale studies on the effects of LULC on water quality are based on a crude classification of land use as agriculture, forested or urban area and do not take into account the seasonal dynamics of land cover and farming practices that characterize the agricultural land use (Dunne, 1979; Hecky *et al.*, 2003; Bagalwa, 2006; de Villiers and Thiart, 2007). The combination of farm, reach and catchment level studies, as applied in this thesis, can help capture more details on the variation of water quality in the landscape.

Nutrient enrichment from agricultural activities is diffuse and difficult to control. Its management requires a combination of appropriate strategies implemented at farm, reach and catchment levels. Similarly, to be effective, studies on the effects of land use on water quality should capture the variation on macro-, meso- and micro-scale landscape levels. River basin (macro) scale studies tend to take a black box approach and incorporate limited information on what is happening within the catchment. Farm level (micro) scale studies run the risk of either overestimating or underestimating the effects of the observed variables, and may not capture catchment processes such as underground flows, re-deposition in nearby fields and in-stream processes that may dilute or concentrate a water variable at the catchment level. On their way from farms to the streams, sediments and nutrients may settle, can be adsorbed on soil particulates or vegetation, transformed into other forms, or leach to the groundwater (N and P). Newly formed or accumulated materials can be mobilized into the watercourse by suspension or dissolution, depending on prevailing catchment characteristics (slope, type of soil/sediments and LULC).

6.3 Surface and groundwater pathways

Catchment water quality is influenced by a number of natural and human induced factors, including frequency and intensity of rainfall, type of LULC, soil type (hydraulic conductivity and erodibility), slope type (slope length and angle), flow paths, biogeochemical processes (sub-surface exchange, adsorption, and denitrification) and farming practices (Sophocleous, 2002; Jennings *et al.*, 2003; Burt and Pinay, 2005; Verhoeven *et al.*, 2006; Lohse et al. 2009; Pärn *et al.*, 2012). Sediment and P transport generally depend on overland flows, while N transport is more related to subsurface flows (Jordan *et al.*, 1997; Pärn *et al.*, 2012). In agricultural catchments, both N and P can be washed into surface waters by overland runoff shortly after the application of fertilizers and manures, or during livestock grazing. Transport of N through groundwater, often in the form of nitrate, is slower (Howden *et al.* 2011). Landscapes with a high hydraulic conductivity increase the recharge of the groundwater, while leaching N, P and other substances (Heathwaite *et al.* 2000). At low surface flow velocity, particulates, along with adsorbed P and N may settle. Natural depressions and river banks can create inundated areas that increase water retention time and sedimentation of particulates, which can also promote other processes such as nutrient uptake by vegetation and microbes, adsorption to soil, and denitrification (van der Lee *et al.*, 2004; Lohse *et al.*, 2009; Pärn *et al.* 2012). Dense vegetation, as found in riparian zones, can resist the movement and transport of water and associated materials (Cooper *et al.*, 2000; Dabney *et al.*, 2001). Nutrients can also be transferred to the groundwater following build-up after fertilization, animal manure loading, or urban and industrial pollution (Jennings *et al.* 2003).

In Migina catchment, the fate of sediment and nutrients is determined by seasonal variation of natural processes and human activities (including farming). During the dry season, build-up of material (sediments and nutrients) occurs following reduced water kinetics. At higher flows, sediments and nutrients are mobilized and move downstream. Due to the high hydraulic conductivity of the ferralitic soils in Migina catchment, rainwater along with dissolved nutrients, infiltrates rapidly to the groundwater. Overland runoff dominates the subsurface flows only during heavy rainfall events (Munyaneza *et al.*, 2012). This runoff is responsible for the transport of the majority sediment and associated P and N particulates (Figure 4 and 5, Chapter 2), while dissolved N is more likely to be transported by subsurface flows. Reaches with natural and human made depressions (pond/reservoir) or grass/forest on river banks increase water retention time and sedimentation of suspended material (Figure 5, Chapter 3).

6.4 Impact of farming

While agricultural intensification is critical to increasing food security and economic development, its environmental impacts may be negative because of the alteration or destruction of wetland ecosystems and biodiversity, and associated impact on their regulating ecosystem services. Land cover changes within the crop growing season can fluctuate from bare soil resulting from hoeing to a dense canopy at the end of the season with intermediate farming stages associated with fertilizer application and weeding. Ploughing makes soil more vulnerable for washout of soil particles and nutrients, whereas a dense established crop, wetland vegetation or ponds may increase the sink function (Dabney, 1996; Dabney at al., 2001).

In Migina catchment, like in many catchments in sub-Saharan Africa, the peak of farming activities coincides with the beginning of the rainy season, when valley bottoms are most vulnerable to the loss of sediment and nutrients. In these periods, loose material from the ploughed soil and built-up material accumulated during the dry season are mobilized and exported at the onset of high rainfall (Uwimana et al., 2017). Highest sediment and nutrient exports were observed during the early stages of the vegetable growing season, when bare soil dominated the crop cover. The later stages of farming were characterized by small exports and sometimes retention (Chapter 3, Chapter 4), which means that farming activities with less soil disturbance, and crops with denser vegetation cover could reduce sediment and nutrient export. More land cover can resist the movement and transport of water and associated materials (Dabney, 1996; Dabney et al., 2001). The effluent discharge from reservoirs and ponds during periods of water release, or when ponds are drained for fish harvesting or dredging, was the major loss pathway of N and P.

Agriculture should minimize sediment and nutrients export at the critical stages of farming, notably during ploughing in rainy season. At the farm level, this can be done through the use of conservation agriculture techniques, such as ploughing with minimal soil disturbance, or minimal use of water. At the valley bottom level, the use of retention ponds and wetland buffer zones to reduce water, sediment and nutrient transport can be considered (Drechsel et al., 1996; Labrière et al., 2015). In the Migina catchment, this would require government intervention in engaging people about the closed-loop agriculture that reduces yield and export of water, sediment and nutrients. The common practice of flooding rice farms to facilitate ploughing, levelling and weeding should be discouraged, as this increases the export of sediments and associated nutrients. Similarly, the practice of discharging the effluent from pond/reservoir drainage and dredging to the nearby water bodies should be replaced by storing the effluent in ponds and reusing sediments in agriculture.

6.5 Regional implications

The finding of this study in the Migina catchment can be applied to other catchments that are influenced by similar drivers of landscape and water quality change. Like many catchments in sub-Saharan Africa, Migina catchment faces problems of high land pressure, steep slopes, heavy rainfall and inadequate land management and soil conservation practices. Research in several other African wetlands, such as Namatala wetland in Uganda (Namaalwa et al., 2013), Nyando wetland in Kenya (Kipkemboi, 2006; Rongoei et al., 2014), wetlands in Illubabur and the swamps of Awash valley in Ethiopia (Gebresllassie et al., 2014) and wetlands in other parts of Africa (Wood and Dixon, 2002; Wood, 2013) show the challenges of different conflicting uses by different stakeholders. The role of these wetlands in sediment and nutrient pathways varies depending on their location and connections to the upstream and downstream waters, and prevailing processes (physical, chemical, and biological) that can concentrate, attenuate or transform sediments, nutrients or other material in water. In the coming years, conversion of wetlands is expected to intensify following population growth and the resulting need for economic growth, and climate change (OECD/FAO, 2016; Ramsar Convention on Wetlands, 2018).

To address the global environmental challenges of population increase, degradation of natural resources and climate change, commonly called "wicked" problems, wetland agricultural development and conservation should be reconciled in all their aspects (Milder et al., 2014; Freeman et al., 2015). The landscape approach has emerged as one of the promising ways to sustainably address the landscape complexity of conflicting functions, services and values to local people and ecological systems (Sayer, 2009; Freeman et al., 2015). This approach conforms to the concept of wise use of wetlands of the Ramsar Convention on Wetlands aimed at sustainable development (Ramsar Convention on Wetlands, 2005; 2018) and the important role of wetlands in sustaining livelihoods (Ramsar, 2005; Ellison and Wood, 2013). In Rwanda, as hillside land is becoming less fertile from over-exploitation, lack of water for irrigation and agricultural inputs, conversion of more natural wetlands to agricultural farms is expected. More actions from different levels of society (livelihood alternatives and government incentives) are needed to facilitate the implementation of the landscape approach.

The landscape approach aims to reconcile livelihood development and conservational goals to achieve the socio-economic development and environmental sustainability (Ellison and Wood, 2013; Sayer et al., 2013). In the valley bottoms of the Migina catchment, a landscape approach could be implemented through balancing multiple wetland uses, like controlled agriculture, aquaculture, extraction of sand, silt and clay and harvesting of natural products (medicine, fodder and material for producing handicrafts), ecotourism and payment for environmental services. Agriculture, aquaculture, sand, silt and clay mining increase income, but often occur at the cost of degradation of hydrology, water quality and biodiversity. In contrast, sustainable harvesting of natural products, ecotourism and other ecological services put less stress on the environment, but may not respond sufficiently to all needs of people in terms of food security and economic development (Dixon and Wood, 2003). Special attention should be taken to ensure the implementation of best farming practices with minimum soil disturbance, optimum agrochemical inputs and less effluent. Subsidiary mechanisms by the government to the compliant farmers should be considered in case the agricultural production is compromised. Similarly, punitive mechanisms for incompliant farmers should be considered. Monetary valuation of wetland ecosystem regulating services could be better used to inform decision making on matters affecting e.g. regulation of water flows, water quality, biodiversity, and mitigation of effects of changing climate.

6.6 Implications to future economic development and climate change

Worldwide, natural land cover (grassland, forest, wetlands) is progressively declining, while agricultural and urban areas are increasing (Naeem *et al.*, 1999; FAO, 2010; FAO, 2011b; van Asselen *et al.*, 2013; WWF, 2014; Ramsar Convention on Wetlands, 2018). Conversion of wetlands to agriculture as a strategy to increase food security is widespread throughout sub Sahara Africa (Schuyt, 2005; Seck *et al.*, 2012; Wood *et al.*, 2013). Wetland conversion is often done without due consideration of the implications for biodiversity and regulating ecosystem services associated with those wetlands. Greater awareness of the links between ecosystem services and agriculture, or requirements for greater regulation is needed to ensure that agriculturally converted wetlands are managed responsibly in ways that minimize cost to the environment.

Wise use of wetland in sub-Saharan Africa can lead to increased food security and economic development, while maintaining ecosystem services (Kotze, 2011; Sakané *et al.*, 2011). Rodenburg *et al.* (2014) argued that if valley bottom wetlands are appropriately selected and developed, more than 90% of wetlands in Africa can be conserved and serve other purposes than agriculture. They proposed a four stage stakeholder-participatory approach (site-selection, land-use planning, water management design and crop management) for maximizing the crop production in small areas while maintaining other important ecosystem functions on the remaining wetlands. Wetland selection and development could be done through a combination of surveys involving GIS and remote sensing, soil depth, soil fertility, water availability and socio-economic variables such as availability of markets, extension services or social customs and stakeholders (Narteh *et al.*, 2007). Approaches like the Working Wetland Potential (WWP) could be used to weigh the suitability of wetlands for agricultural development in terms of ecological potential, social and economic importance against the environmental and socio-economic risks involved in the actual development (McCartney and Houghton-Carr, 2009).

In Rwanda, in an attempt at proper management of the wetlands and water quality, two important orders on the protection of wetlands, shores of lakes and rivers were enacted. One of the orders (Prime Minister's Order No 006/03 of 30/01/2017) classified 935 swamps in the categories "Use under specific conditions", "Use without specific conditions", or "Full Protection" (Government of Rwanda, 2017). In swamps under full protection, it is prohibited to carry out any activities, except those related to research and science. Only about 40 swamps, mainly found in the national parks or upstream of intense human activities (urban areas and agriculture) are in the list of full protected wetlands. The remaining wetlands (about 900), located mostly in agricultural catchments, are classified as "Use under specific conditions" and "Use without specific conditions". Apart from being reserved for tourism, biodiversity conservation and as sources of water, the spatial arrangement of the full protected wetlands in national parks or upstream of the exploited catchments make them less usable for controlling floods or removing sediment and nutrients from the agricultural and urban areas. To protect and support the important ecosystem services of many wetlands, it is important to reconsider the list by strategically adding to the list of full protected wetlands some wetlands found downstream of urban and agricultural areas. Another ministerial order designated strips of 50, 10 and 5 m of land from the shore of lakes, big rivers and small rivers, respectively, as protected buffer zone (Government of Rwanda, 2010). The enforcement of this order would increase the presence of natural vegetation in the riparian and lake buffer zone and could improve water quality in streams, rivers and lakes.

6.7 Recommendations for future research

This study assessed the effects of conversion of natural wetlands to agriculture on the water quality and retention of sediments, N and P. However, the study did not capture all short-term variations (daily and hourly) for the considered water variables. Some processes like surface-groundwater interaction, denitrification and ammonia volatilization were not satisfactorily covered, although they can affect retention in the catchment considerably. Further studies should focus on the short term (hourly and daily) variations of sediment and nutrients for the dry period (build-up period) and heavy rainfall events (washout period) and the interaction between surface and groundwater, as well as other processes like denitrification and ammonia volatilization. Further studies should also identify the appropriate arrangement (including construction, restoration or rehabilitation) of different land uses such as water reservoirs, ponds, and natural land cover (grassland and forest) to maximize the retention of sediment and nutrients in the landscape. While this study was based on field observation, satellite images, maps and photographs to assess monthly changes in LULC and manual collection of water variables, further studies should consider using aerial photography (drones), remote sensing and automated samplers to increase the extent and precision of collected LULC and water variables. There is a need for more routine and long-term monitoring of water quality, results of which could be used for adaptive management. With more robust datasets on water variables and meteorological variables (rainfall, temperature, relative humidity, wind speed and solar radiation), models like for example the Soil and Water Assessment Tool (SWAT/GIS) and Agricultural Policy/Environmental Extender (APEX) could be considered to determine the relative sizes and combinations of different LULC types for optimal crop production and environmental protection in the context of higher pressure on land similar to Migina catchment.

7. References

Adekola, O., Whanda, S., Ogwu, F., 2012. Assessment of Policies and Legislation that Affect Management of Wetlands in Nigeria. *Wetlands* **32,** 665–677. doi:10.1007/ s13157-012-0299-3

AfDB/OECD, 2007. *African Economic Outlook-Rwanda.* Available at: http:// www.oecd.org/2009.

APHA, 2005. *Standard methods for the examination of water and wastewater,* 21st edition. American Public Health Association, Washington DC, USA.

ASWM, 2005. *Common questions: Wetlands, climate change, and carbon sequestering.* Association of State Wetland Managers Inc., USA. Available at http:// www.aswm.org/2009.

ATTRA, 2009. *Agriculture, Climate Change and Carbon Sequestration.* National Sustainable Agriculture Information Service / NCAT Program Specialists. 2009 NCAT. Available online on www.ncat.attra.org/2009.

Avnimelech, Y., 2009. Biofloc Technology - A Practical Guide Book, first ed. The World Aquaculture Society, Baton Rouge, Louisiana, pp. 182.Barszczewski, J., Kaca, E., 2012. Water retention in ponds and the improvement of its quality during carp production. *Journal of Water and Land Development* **17,** 31–38.

Avnimelech, Y., 1998. Minimal discharge from intensive fishponds. *World Aquaculture* **9,** 32-37.

Bae, H.K., 2013. Changes of river's water quality responded to rainfall events. *Environment and Ecology Research* **1,** 21-25.

Bagalwa, M., 2006. The impact of land use on water quality of the Lwiro River, Democratic Republic of Congo, Central Africa. *African Journal of Aquatic Science* **31,** 137-143.

Banas, D., Masson, G., Leglize, L. and Pihan, J.C., 2002. Discharge of sediments, nitrogen (N) and phosphorus (P) during the emptying of extensive fishponds: effect of rain-fall and management practices. *Hydrobiologia* **472,** 29-38. https:// doi.org/10.1023/A:1016360915185

Bartley, R., Speirs, W., 2010. *Review and summary of constituent concentration data from Australia for use in catchment water quality models.* Technical Report, eWater Cooperative Research Centre, Bruce, Australia.

Bolstad, P.V., Swank, W.T., 1997. Cumulative impacts of landuse on water quality in a southern Appalachian watershed. *Journal of the American Water Resources Association* **33,** 519-533.

Bouman, B.A.M, Peng, S., Castañeda, A.R, Visperas, R.M., 2005. Yield and water use of irrigated tropical aerobic rice systems. *Agricultural Water Management* **74,** 87–105.

Boyd, C.E., Tucker, C.S., 1995. Sustainability of channel catfish farming. *World Aquaculture* **26,** 45-53.

Boyd, C.E., Tucker, C.S., 1998. *Pond water quality management*. Kluwer Academic Public., Nowell, MA.

Bullock, A., Acreman, M., 2003. The role of wetlands in the hydrological cycle. *Hydrology and Earth Systems Science* **7**, 358-389.

Burt, T.P., Pinay, G., 2005. Linking hydrology and biogeochemistry in complex landscapes. *Progress in Physical Geography* **29**, 297-316.

Cao, Z.H., Zhang, H.C., 2004. Phosphorus losses to water from lowland rice fields under rice-wheat double cropping system in the Tai Lake region. *Environmental Geochemistry and Health* **26**, 229-36.

Carroll, K.P., Rose, S., Peters, N.E., 2007. *Concentration/discharge hysteresis analysis of storm events at the Panola mountain research watershed, Georgia, USA*. Proceedings of the Georgia Water Resources Conference, March 27-29, 2007. University of Georgia, Athens, GA, USA.

Cheruiyot, C., Muhandiki, V., 2014. Review of Estimation of Pollution Load to Lake Victoria. *Journal of Environment and Earth Science* **4**, 113-120.

Christen, E., Jayewardene, N., 2005. *Water use efficiency and water use productivity in irrigation*. In: Meyer WS. (ed.) The irrigation industry in the Murray and Murrumbidgee Basins. Technical Report 03/05, CRC for Irrigation Futures; June 2005, pp 89-92.

Ciria, M.P., Solano, M.L., Soriano, P., 2005. Role of Macrophyte Typha latifolia in a Constructed Wetland for Wastewater Treatment and Assessment of its Potential as a Biomass Fuel. *Biosystems Engineering* **92**, 535-544.

Cooper, C.M., Shields, F.D., Testa, J.S., Knight, S.S., 2000. Sediment retention and water quality enhancement in disturbed watersheds. *International Journal of Sediment Research* **15**, 121-134.

Coulliette, A.D., Noble, R.T., 2008. Impacts of rainfall on the water quality of the Newport River Estuary (Eastern North Carolina, USA). *Journal of Water and Health* **6(4)**, 473-482.

Dabney, S.M., 1996. Cover crop impacts on watershed hydrology. *Soil and Water Conservation* **53**, 207–213.

Dabney, S.M., Delgado, J.A., Reeves, D.W., 2001. Using winter cover crops to improve soil and water quality. *Communications in Soil Science and Plant Analysis* **32**, 1221-1250..

Davari, M., Ram, M., Tewari, J., Kaushish, S., 2010. Impact of agricultural practice on ecosystem services. *International journal of Agronomy and Plant Production* 1(1), 11-23.

Davidson, N.C., 2014. How much wetland the world has lost? Long-term and recent trends in global wetland area. *Marine and Freshwater Research* **65**, 934-941.

Day, W., Audsley, E., Frost, A.R., 2008. An engineering approach to modelling, decision support and control for sustainable systems. *Philosophical Transaction of the Royal Society B*, **363**, 527-541 de Villiers, S., Thiart C., 2007. The nutrient status of South African rivers: concentrations, trends and fluxes from the 1970s to 2005. *South African Journal of Science* **103**, 343-349.

Demars, B.O.L., Harper, D.M., Pitt, J.A., Slaughter, R., 2005. Impact of phosphorus control measures on in-river phosphorus retention associated with point source pollution. *Hydrology and Earth System Sciences Discussions* **2**, 37–72.

Denny, P., 1997. Implementation of constructed wetlands in developing countries. *Water Science and Technology* **35**, 27-34.

Díaz, R.J., Rabalais, N.N., Breitburg, D.L., 2012. *Agriculture's Impact on Aquaculture: Hypoxia and Eutrophication in Marine Waters*. OECD Report 2012. Available on http://www.oecd.org/tad/sustainable-agriculture/49841630.

Dixon, A.B., Wood, A.P., 2003. Wetland cultivation and hydrological management in eastern Africa: Matching community and hydrological needs through sustainable wetland use. *Natural Resources Forum* **27**, 117–129.

Dobermann, A., Witt, C., 2004. *The evolution of site-specific nutrient management in irrigated rice systems of Asia.* pp. 76-100, In: Increasing productivity of intensive rice systems through site-specific nutrient management. Enfield, (USA) and Los Baños (Philippines): Science Publishers, Inc., and International Rice Research Institute.

Drechsel P., Steiner K.G., Hagedorn F., 1996. A review on the potential of improved fallows and green manure in Rwanda. *Agroforestry Systems* **33**, 109-136.

Drexler, J.Z., Anderson, F.E., Snyder, R.L., 2008. Evapotranspiration rates and crop coefficients for a restored marsh in the Sacramento–San Joaquin Delta, California, USA. *Hydrological Processes* **22**, 725–735. doi: 10.1002/hyp.6650

Dunne, T. , 1979. Sediment yield and land use in tropical catchments. *Journal of Hydrology* **42**, 281-300.

Ellison, J., Wood, A., 2013. *Functional Landscape Approach: Baseline Assessment of Four Sites in Karonga District for the DISCOVER Project*. Technical Report July 2013, Wetland Action. doi: 10.13140/RG.2.2.20972.36485

Emerson, K., Russo, R.C. Lund R.E., Thurston, R.V., 1975. Aqueous ammonia equilibrium calculations: effect of pH and temperature. *Journal of the Fisheries Research Board of Canada* **32**, 2379-2383.

Everitt, B.S., Hothorn, T., 2010. *A handbook of statistical analyses using R*, 2nd edition. CRC Press (Taylor and Francis Group), Boca Raton, FL, USA.

FAO, 1998. Crop evapotranspiration - guidelines for computing crop requirements. *FAO Irrigation and Drainage Paper 56*. Food and Agriculture Organization of the UN, Rome. Available at: www.fao.org/docrep/2014

FAO, 2010. Global Forest Resources Assessment Main Report. *FAO Forestry Paper 163*. Food and Agriculture Organization of the UN, Rome.

FAO, 2011a. Climate change, water and food security. *FAO Water Reports 36*. Food and Agriculture Organization of the United Nations, Rome, Italy.

FAO, 2011b. *The state of the world's land and water resources for food and agriculture (SOLAW) – managing systems at risk*. Rome: Food and Agriculture Organization of the United Nations, and London: Earthscan.

Fillery, I.R.P., Vleck, P.L.G., 1986. Reappraisal of the significance of ammonia volatilization as a N loss mechanism in flooded rice fields. *Fertilizer Research* **9**, 79-98.

Freeman, O.E., Duguma, L.A., Minang, P.A., 2015. Operationalizing the integrated landscape approach in practice. *Ecology and Society* **20**(1), 24. http://www.ecologyandsociety.org/vol20/iss1/art24/

Gebresllassie, H., , Gashaw, T., Mehari, A., 2014. Wetland Degradation in Ethiopia: Causes, Consequences and Remedies. *Journal of Environment and Earth Science* **4**, 40-48.

Giertz, S., Junge, B., Diekkrüger, B., 2005. Assessing the effects of land use change on soil physical properties and hydrological processes in the sub-humid tropical environment of West Africa. *Physics and Chemistry of the Earth* **30**, 485–496.

Giga, J.V., Uchrin, C.G., 1990. Laboratory and in situ sediment oxygen demand determinations for a Passaic river (NJ) case study: Part 1. *Journal of Environmental Science and Health* **25**, 833-845.

Government of Rwanda, 2010. Ministerial Order n°007/16.01 of 15/07/2010 Determining the length of land on shores of lakes and rivers transferred to public property. *Official Gazette of the Republic of Rwanda* **37** (13/09/2010).

Government of Rwanda, 2017. Prime Minister's Order No 006/03 of 30/01/2017 drawing up a list of swamp lands, their characteristics and boundaries and determining modalities of their use, development and management. *Official Gazette of the Republic of Rwanda* **7** (13/02/2017).

Green, B.W., Boyd, C.E., 1995. Chemical budget for organically fertilized fishponds in the dry tropic. *Journal of World Aquaculture Society* **26**, 284-296.

Gross, A., Boyd, C.E., Wood, C.W., 1999. Ammonia volatilization from freshwater fishponds. *Journal of Environmental Quality* **28**, 793-797.

Haefele, S.M., Wopereis, M.C.S., Ndiaye, M.K., Barro, S.E., Ould Isselmo, M., 2003. Internal nutrient efficiencies, fertilizer recovery rates and indigenous nutrient supply of irrigated lowland rice in Sahelian West Africa. *Field Crops Research* **80**, 19-32.

Hargreaves, J.A., 1998. Nitrogen biogeochemistry of aquaculture ponds. *Aquaculture* **166**, 181–212.

Haygarth, P.M., Jarvis, S.C., 1999. Transfer of phosphorus from agricultural soils. *Advances in Agronomy* **66**, 196–249.

Heathwaite, L., Sharpley, A., Gburek, W., 2000. A conceptual approach for integrating phosphorus and nitrogen management at watershed scales. *Journal of Environmental Quality* **29**, 158-166.

Hecky, R.E., Harvey, A., Bootsma, H.A., Kingdon, M.L., 2003. Impact of land use on sediment and nutrient yields to Lake Malawi/Nyasa (Africa). *Journal of Great Lakes Research* **29**, 139–158.

Henao, J., Baanante, C., 1999. *Estimating rates of nutrient depletion in soils of agriculture lands in Africa*. Muscle Shoals, US: International Fertilizer Development Center.

Henry, C.P., Amoros, C., 1995. Restoration ecology of riverine wetlands: I. A scientific base. *Environmental Management* **19**, 891-902.

Hilderbrand, R.H., Watts, A.C., Randle, A.M., 2005. The myths of restoration ecology. *Ecology and Society* **10**(1), 19. http://www.ecologyandsociety.org/vol10/iss1/art19/

Hirsch, R.M., 2012. Flux of nitrogen, phosphorus, and suspended sediment from the Susquehanna River Basin to the Chesapeake Bay during Tropical Storm Lee, September 2011, as an indicator of the effects of reservoir sedimentation on water quality. *US Geological Survey Scientific Investigations Report* **2012–5185,** 17 p.

Howden, N.J.K., Burt, T.P., Worrall, F., Mathias, S.A., Whelan, M.J., 2011. Nitrate pollution in intensively farmed regions: What are the prospects for sustaining high quality groundwater. *Water Resources Research* **47**(6), W00L02. doi:10.1029/ 2011WR010843.

Hu, Z.Q., Wu, S., Ji, C., Zou, J.W., Zhou, Q.S., Liu, S.W., 2015. A comparison of methane emissions following rice paddies conversion to crab-fish farming wetlands in southeast China. *Environmental Science and Pollution Research* **23**(2), 1505-1515. doi:10.1007/s11356-015-5383-9

Husnain, H., Masunaga, T., Wakatsuki, T., 2010. Field assessment of nutrient balance under intensive rice-farming systems, and its effects on the sustainability of rice production in Java Island, Indonesia. *Agricultural, Food, and Environmental Sciences* **4**, 11p.

ICARDA, 2001. *Soil and Plant Analysis Laboratory Manual*, Second Edition. International Center for Agricultural Research in the Dry Areas (ICARDA) and National Agricultural Research Center (NARC). Pakistan: Islamabad.

Ijaz, A., Khan, F., Bhatti, A.U., 2007. Soil and nutrient losses by water erosion under monocropping and legume inter-cropping on slopping land. *Pakistan Journal of Agricultural Research* **20**, 161-166.

Jansky, L., Chandran, R., 2004. Climate change and sustainable land management: Focus on erosive land degradation. *Journal of World Association of Soil and Water Conservation* **4**, 17–29.

Jason, T.M., Vellidis, G., Lowrance, R., Pringle, C.M., 2009. High sediment oxygen demand within an instream swamp in southern Georgia: implications for low dissolved oxygen levels in coastal blackwater streams. *Journal of American Water Resources Association* **45**, 1493-1507.

Jennings, E., Irvine, K., Mills, P., Jordan, P., Jensen, J-P., Søndergaard, M., Barr, A., Glasgow, G., 2003. *Eutrophication from agricultural sources, seasonal patterns and effects of phosphorus*. Final report to the Irish EPA (Environmental RTDI Programme 2000-2006, 2000-LS-2.1.7-M2), 77 p. Environmental Protection Agency, Johnstown Castle, Co. Wexford, Ireland.

Jiménez-Montealegre, R., Verdegem, M., Zamora J.E., Verreth, J.A.V., 2002. Organic matter sedimentation and resuspension in tilapia (Oreochromis niloticus) ponds during a production cycle. *Aquacultural Engineering* **26**, 1–12.

Johnson, L.B., Richards, C., Host, G.E., Arthur, J.W., 1997. Landscape influences on water chemistry in midwestern stream ecosystems. *Freshwater Biology* **37**, 193–208.

Johnston, C.A., 1991. Sediment and nutrient retention by freshwater wetlands: effects on surface water quality. *Critical Reviews in Environmental Control* **21**, 491–565.

Jordan, T.E., Correll, D.L., Weller, D.E., 1997. Relating nutrient discharges from watersheds to land use and streamflow variability. *Water Resources Research* **33**, 2579-2590.

Kadlec, R.H., Wallace, S.D., 2009. *Treatment Wetlands*, 2nd ed. CRC Press, Taylor & Francis Group, Boca Raton, Florida, USA.

Kingdon, M.J., Bootsma, H.A., Mwita, J., Mwichande, B., Hecky, R.E., 1999. *Discharge and water quality*. Chapter 2, in: Bootsma H.A., Hecky R.E. (Eds.) Water quality report, Lake Malawi/Nyasa Biodiversity Conservation Project, pp. 29-69. Southern African Development Community (SADC) and Global Environmental Facility (GEF), Senga Bay, Malawi.

Kipkemboi, J., 2006. *Fingerponds: seasonal integrated aquaculture in East African freshwater wetlands Exploring their potential for wise use strategies*. PhD thesis, Wageningen University and UNESCO-IHE Institute for Water Education. Taylor & Francis/Balkema, Leiden, The Netherlands.

Kipkemboi, J., Luoga, H.P., van Dam, A.A., Denny, P., 2010. Assessment of Nutrient flows and sustainability in smallholder wetland-Terrestrial Farming Systems at the Shores of Lake Victoria, Kenya using Ecopath. *Egerton Journal of Science and Technology* **10**, 58-84.

Klemedtsson L., Hansson G., Mosier A. (1990) *The use of acetylene for the quantification of N_2 and N_2O production from biological processes in soil*. In: Revsbech N.P., Sørensen J. (eds) Denitrification in Soil and Sediment. Federation of European Microbiological Societies Symposium Series, volume 56. Springer, Boston, MA.

Knowles R. (1990) *Acetylene inhibition technique: development, advantages, and potential problems*. In: Revsbech N.P., Sørensen J. (eds) Denitrification in Soil and Sediment. Federation of European Microbiological Societies Symposium Series, volume 56. Springer, Boston, MA.

Konnerup, D., Trang, N.T.D., Brix, H., 2011. Treatment of fishpond water by recirculating horizontal and vertical flow constructed wetlands in the tropics. *Aquaculture* **313,** 57-64.

Kotir, J.H., 2011. Climate change and variability in sub-Saharan Africa: a review of current and future trends and impacts on agriculture and food security. *Environment, Development and Sustainability* **13,** 587–605.

Kotze, D.C., 2011. The application of a framework for assessing ecological condition and sustainability of use to three wetlands in Malawi. *Wetland Ecology and Management* **19,** 507–520.

Krupnik, T.J, Six, J., Ladha, J.K., Paine, M.J., van Kessel, C., 2004. *An assessment of fertilizer nitrogen recovery efficiency by grain crops.* In: Agriculture and the nitrogen cycle: assessing the impacts of fertilizer use on food production and the environment (editors: A.R. Mosier, J.K. Syers and J.R. Freney), pp. 193-207. Island Press, Washington, USA.

Kufel, L., 2012. Are fishponds really a trap for nutrients? – a critical comment on some papers presenting such a view. *Journal of Water and Land Development* **17,** 39-44.

Labrière, N., Locatelli, B., Laumonier, Y., Freycon, V., Bernoux, M., 2015. Soil erosion in the humid tropics: A systematic quantitative review. *Agriculture, Ecosystems & Environment* **203,** 127-139.

Ladha, J.K., Boonkerd, N., 1988. *Biological nitrogen fixation by heterotrophic and phototrophic bacteria in association with straw.* In: Proceedings of the First International Symposium on Paddy Soil Fertility, Chiangmai, Thailand, 6-13 December 1988. pp 173-187. ISSS, Paddy Soil Fertility Group.

Lohse, K.A., Brooks, P.D., McIntosh, J.C., Meixner, T., Huxman, T.E., 2009. Interactions between biogeochemistry and hydrologic systems. *Annual Review of Environment and Resources* **34,** 65-96.

Mangiafico, S.S., 2016. *Summary and analysis of extension education progam evaluation in R, version 1.6.16.* Rutgers Cooperative Extension, New Brunswick, NJ, USA. Available as: rcompanion.org/documents/RHandbook ProgramEvaluation.pdf

McCartney, M., Rebelo, L-M., Senaratna Sellamuttu, S., de Silva, S., 2010. *Wetlands, agriculture and poverty reduction.* IWMI Research Report 137. International Water Management Institute, Colombo, Sri Lanka.

McCartney, M.P., Houghton-Carr, H.A., 2009. Working Wetland Potential: An index to guide the sustainable development of African wetlands. *Natural Resources Forum* **33(2),** 99-110.

McClain, E.M., Boyer, W.E., Dent, C.L., Gergel, E.S., Grimm, B.N., Groffman, M.P., Hart, C.S., Harvey, W.J., Johnston, A.C., Mayorga, E., McDowell, H.W., Pinay, G., 2003. Biochemical hot spot and hot moments at the interface of terrestrial and aquatic ecosystems. *Ecosystems* **6,** 301-312.

MEA, 2005. *Millennium Ecosystem Assessment. Ecosystems and Human Well-being: Synthesis.* Island Press, Washington, DC.

Milder, J.C., Hart, A.K., Dobie, P., Minai, J., Zaleski, C., 2014. Integrated landscape initiatives for African agriculture, development, and conservation: a region-wide assessment. *World Development* **54,** 68-80. http://dx.doi.org/10.1016/j.worlddev.2013.07.006

Milliman, J.D., Syvitski, J.P.M., 1992. Geomorphic/tectonic control of sediment discharge to the ocean: the importance of small mountainous rivers. *Journal of Geology* **100,** 525–544.

MINAGRI, 2009. *Strategic Plan for the Transformation of Agriculture in Rwanda – Phase II* (PSTA II). Final report. Ministry of Agriculture and Animal Resources, Republic of Rwanda, p. 114. http://www.https://www.gafspfund.org/ (accessed 01.10.15).

MINAGRI, 2010a. *Rwanda Irrigation Master Plan.* Ministry of Agriculture and Animal Resources, Republic of Rwanda, p. 229. https://www.gafspfund.org/2015/ (accessed 01.10.15).

MINAGRI, 2010b. *Enabling Self Sufficiency and Competitiveness of Rwanda Rice: Issues and Policy Options.* Ministry of Agriculture and Animal Resources, Republic of Rwanda, Kigali. http://www.minagri.gov.rw/fileadmin/user_upload/documents/agridocs/Rwa_RicePolicyReport.pdf (accessed 06.12.17)

MINAGRI, 2011. *Master Plan for Fisheries and Fish farming in Rwanda.* Ministry of Agriculture and Animal Resources, Republic of Rwanda, Kigali. http://www.minagri.gov.rw/fileadmin/user_upload/documents/STRAT.PLC/Fisheries_and_Fish_Farming_Master Plan_2__1.pdf) (accessed 06.02.17)

MINAGRI, 2013a. *National rice development strategy (Period 2011-2018).* Ministry of Agriculture and Animal Resources, Kigali, Rwanda. Available at http://www.minagri.gov.rw (accessed 27.03.14)

MINAGRI, 2013b. *Strategic Plan for the Transformation of Agriculture in Rwanda Phase III* (PSTA III), Final report. Ministry of Agriculture and Animal Resources, Republic of Rwanda, Kigali.

MINAGRI, 2014. *National Fertilizer Policy.* Ministry of Agriculture and Animal Resources, Republic of Rwanda, Kigali. http://www. minagri.gov.rw/ (accessed 11.10.15).

Ministère des Ressources Naturelles et Service Geologique, 1981. *Carte lithologique du Rwanda.* Institut Géographique National de Belgique, Brussels.

MINITERE, 2004. *Sectorial Policy on Water and Sanitation.* Ministry of Lands, Environment and Forests, Water and Natural Resources, Republic of Rwanda, Kigali.

Mitsch, W.J., Gosselink, J.G., 2000. *Wetlands.* 3rd Edition, John Wiley and Sons Inc, New York, USA.

Mitsch, W.J., Wu, X., Nairn, R.W., Weihe, P.E., Wang, N., Deal, R., Boucher, C.E., 1998. Creating and restoring wetlands. *BioScience* **48,** 1019-1030.

Mitsch, W.J., Zhang, L., Anderson, C.J., Altor, A.E., Hernandez, M.E., 2005. Creating riverine wetlands: Ecological succession nutrient retention, and pulsing effects. *Ecological Engineering* **25,** 510–527.

Mkanda, F.X., 2002. Contribution by farmer's survival strategies to soil erosion strategies in the Linthipe River catchment: implications for biodiversity conservation in Lake Malawi/Nyasa. *Biodiversity & Conservation* **11,** 1327–1359.

Mohanty, R.K., Ambast, S.K., Panigrahi, P., Mandal, K.G., 2018. Water quality suitability and water use indices: Useful management tools in coastal aquaculture of Litopenaeus vannamei. *Aquaculture* **485,** 210–219.

Moomaw, W.R., Chmura, G.L., Davies, G.T., Finlayson, C.M., Middleton, B.A., Natali, S.M., Perry, J.E., Roulet, N., Sutton-Grier, A.E., 2018. Wetlands in a changing climate: science, policy and management. *Wetlands* **38,** 183-205. https://doi.org/10.1007/s13157-018-1023-8

Muendo, P.N., Verdegem, M.C., Stoorvogel, J.J., Milstein, A., Gamal, E.N., Duc, P.M., Verreth, J.A.J., 2014. Sediment accumulation in fish ponds; its potential for agricultural use. *International Journal of Fisheries and Aquatic Studies* **1,** 228-241.

Munyaneza, O., 2014. *Space-time variation of hydrological processes and water resources in Rwanda - focus on the Migina catchment*. PhD-thesis, UNESCO-IHE Institute for Water Education, Delft, The Netherlands.

Munyaneza, O., Mukubwa, A., Maskey S., Uhlenbrook, S., Wenninger, J., 2014. Assessment of surface water resources availability using catchment modelling and the results of tracer studies in the mesoscale Migina Catchment, Rwanda. *Hydrology and Earth Systems Science* **18,** 5289–5301.

Munyaneza, O., Ufiteyezu, F., Wali, U.G., Uhlenbrook, S., 2011. A simple Method to Predict River Flows in the Agricultural Migina Catchment in Rwanda. *Nile Water Science and Engineering Journal* **4,** 24-36.

Munyaneza, O., Wenninger, J., Uhlenbrook, S., 2012. Identification of runoff generation processes using hydrometric and tracer methods in a meso-scale catchment in Rwanda. *Hydrology and Earth Systems Science* **16,** 1991–2004.

Murekatete, E., 2013. *Controls of Denitrification in Agricultural Soils, Wetlands, and Fish Ponds in the Migina Catchment, Rwanda*. MSc-thesis, UNESCO-IHE Institute for Water Education, Delft, the Netherlands.

Mwanja, W.W., Signa, D., Eshete, D., 2011. *Fisheries and Aquaculture Sector Review for Eastern Africa*. SFE Technical Documents Series, Food and Agriculture Organization of the United Nations, Sub Regional Officer for East Africa - Addis Ababa.

Naeem, S., Chapin III, C.F.S., Costanza, R., Ehrlich, P.R., Golley, F.B., Hooper, D.U., Lawton, J.H., Neill, R.V.O., Mooney, H.A., Sala, O.E., Symstad, A.J., Tilman, D., 1999. *Biodiversity and Ecosystem Functioning: Maintaining Natural Life Support Processes.* Issues in Ecology 4, Ecological Society of America. available at: https://www.esa.org/esa/wp-content/uploads/2013/03/issue4.pdf

Namaalwa, S., van Dam, A.A., Funk, A., Ajie, G.S., Kaggwa, R.C., 2013. Characterization of the drivers, pressures, ecosystem functions and services of Namatala wetland, Uganda, *Environmental Science & Policy* **34**, 44-57. http://dx.doi.org/10.1016/ j.envsci.2013.01.002

Narteh, L.T., Moussa, M., Otoo, E., Andah, W.E.I., Asubonteng, K.O., 2007. Evaluating inland valley agro-ecosystems in Ghana using a multi-scale characterization approach. *Ghana Journal of Agricultural Science* **40**, 141–157.

Nhan, D. K., Verdegem, M. C. J., Milstein, A., Verreth, J. A.V., 2008. Water and nutrient budgets of ponds in integrated agriculture–aquaculture systems in the Mekong Delta, Vietnam. *Aquaculture Research* **39**, 1216-1228.

Nhapi, I., Wali, U.G., Uwonkunda, B.K., Nsengimana, H., Banadda, N., Kimwaga, R., 2011. Assessment of water pollution levels in the Nyabugogo Catchment, Rwanda. *Open Environmental Engineering Journal* **4**, 40-53.

NISR, 2008. *Economic Statistics/Environment.* National Institute of Statistics of Rwanda, Kigali. Available at http://statistics.gov.rw/2009.

NISR, 2014. *Fourth Population and Housing Census, Rwanda, 2012.* Thematic Report Population size, structure and distribution January 2014. National Institute of Statistics of Rwanda, Kigali.

OECD/FAO, 2016. *Agriculture in Sub-Saharan Africa: prospects and challenges for the next decade.* In: OECD-FAO Agricultural Outlook 2016-2025. Paris: OECD Publishing.

Ohio EPA, 2004. *Biogeochemical and Hydrological Investigations of Natural and Mitigation Wetlands: Part 5. Integrated Wetland Assessment Program 2004.* Wetland Ecology Group/Environmental Protection Agency/Division of Surface Water. Ohio EPA Technical Report WET/2004-5.

Pang, X.B., Mu, Y.J., Lee, X.Q., Fang, S.X., Yuan, J., Huang, D.K., 2009. Nitric oxides and nitrous oxide fluxes from typical vegetables cropland in China: effects of canopy, soil properties and field management. *Atmospheric Environment* **43**, 2571–2578.

Pant, H.K., Rechcigl, J.E., Adjei, M.B., 2003. Carbon sequestration in wetlands: Concept and estimation. *Food, Agriculture and Environment* **1(2)**, 308-313.

Pärn, J., Pinay, G., Mander, Ü., 2012. Indicators of nutrients transport from agricultural catchments under temperate climate: a review. *Ecological Indicators* **22**, 4-15.

Peng, S.Z., Yang, S.H., Xu, J.Z., Luo, Y.F. and Hou, H.J., 2011. Nitrogen and phosphorus leaching losses from paddy fields with different water and nitrogen managements. *Paddy and Water Environment* **9**, 333-342. doi 10.1007/s10333-010-0246-y

Phong, L.T., Stoorvogel, J. J., van Mensvoort, M.E.F., Udo, H.M.J., 2011. Modeling the soil nutrient balance of integrated agriculture-aquaculture systems in the Mekong Delta, Vietnam. *Nutrient Cycling in Agroecosystems* **90,** 33–49.

Pinheiro, J., Bates, D., DebRoy, S., Sarkar, D., R Core Team, 2017. *nlme: Linear and Nonlinear Mixed Effects Models*. R package version 3.1-131.1, https://CRAN.R-project.org/ package=nlme.

Pokorný, J., Květ, J., 2016. *Fishponds of the Czech Republic*. In: Finlayson, C., Milton, G., Prentice, R., Davidson, N., (eds) The Wetland Book. Springer, Dordrecht. https://doi.org/10.1007/978-94-007-6173-5_208-2

R Core Team, 2016. *R: A language and environment for statistical computing*. R Foundation for Statistical Computing, Vienna, Austria. https://www.R-project.org/.

Rădoane, M., Rădoane, N., 2005. Dams, sediment sources and reservoir silting in Romania. *Geomorphology* **71,** 112–125.

Rahman, M.M., Yakupitiyage, A., 2003. Use of fishpond sediment for sustainable aquaculture-agriculture Farming. *International Journal of Sustainable Development and Planning* **1,** 192–202.

Ramsar Convention on Wetlands, 2005. *A Conceptual Framework for the wise use of wetlands and the maintenance of their ecological character*. Resolution IX.1, Annex A, 9th Meeting of the Conference of the Parties to the Convention on Wetlands (Ramsar, Iran, 1971) Kampala, Uganda, 8-15 November 2005.

Ramsar Convention on Wetlands, 2018. *Global Wetland Outlook: State of the World's Wetlands and their Services to People*. Gland, Switzerland: Ramsar Convention Secretariat. Available at: https://www.global-wetland-outlook.ramsar.org/.

Rebelo, LM., McCartney, M.P., Finlayson, C.M., 2010. Wetlands of Sub-Saharan Africa: distribution and contribution of agriculture to livelihoods. *Wetlands Ecology and Management* **18,** 557-572.

Reddy, K.R., Delaune, R.D., 2008. *Biogeochemistry of wetlands: science and applications*. CRC Press/Taylor and Francis Group, Boca Raton, FL

RNRA, 2012. *Water quality monitoring in Rwanda. Report III: October-November 2012*. Rwanda Natural Resources Authority, Kigali, Rwanda. http://rnra.rw/ fileadmin/fileadmin/user_upload/Water_Quality_monitoring (accessed 03.03.16)

Rodenburg, J., Sander, S.J., Kiepe, P., Narteh, L.T., Dogbe, W., Wopereis, M.C.S., 2014. Sustainable rice production in African inland valleys: Seizing regional potentials through local approaches. *Agricultural Systems* **123,** 1–11.

Roger, P.A., Ladha, J.K., 1992. Biological N fixation in wetland rice fields: estimation and contribution to nitrogen balance. *Plant and Soil* **141,** 41-55.

Roger, P.A., Reddy, P.M., Remulla-Jimenez, R., 1988. *Photodependent acetylene reducing activity (ARA) in ricefields under various fertilizer and biofertilizer management*, p. 827, In: Nitrogen Fixation: Hundred Years After (ed. by Bothe H., de Bruijn F.J., Newton, W.E.). Gustav Fischer, Stuttgart, New York.

Römkens, M.J., Helming, K., Prasad, S.N., 2002. Soil erosion under different rainfall intensities, surface roughness, and soil water regimes. *Catena* **46**, 103-123.

Rongoei, P.J.K., Kipkemboi, J., Okeyo-Owuor, J.B., van Dam, A.A., 2013. Ecosystem services and drivers of change in Nyando floodplain wetland, Kenya. *African Journal of Environmental Science and Technology* **7(5)**, 274-291. doi:10.5897/AJEST12.224

Ruttner-Kolisko, A., 1977. Comparison of various sampling techniques, and results of repeated sampling of planktonic rotifers. *Archiv für Hydrobiologie – Beiheft Ergebnisse der Limnologie* **8**, 13-18.

Sakané, N., Alvarez, M., Becker, M., Böhme, B., Handa, C., Kamiri, H.W., Langensiepen, M., Menz, G., Misana, S., Mogha, N.G., Möseler, B.M., Mwita, E.J., Oyieke, H.A., van Wijk, M.T., 2011. Classification, characterisation, and use of small wetlands in East Africa. *Wetlands* **31**, 1103–1116.

Sayer, J.A., 2009. Reconciling conservation and development: Are landscapes the answer? *Biotropica* **41(6)**, 649–652.

Scanlon, B.R., Jolly, I., Sophocleous, M., Zhang, L., 2007. Global impacts of conversions from natural to agricultural ecosystems on water resources: quantity versus quality. *Water Resources Research* **43**, W03437. doi:10.1029/2006WR005486.

Schnier, H.F., 1995. *Significance of timing and method of N fertilizer application for the N-use efficiency in flooded tropical rice.* In: Ahmad, N. (ed.) Nitrogen Economy in Tropical Soils. Developments in Plant and Soil Sciences, vol 69. Springer, Dordrecht.

Schulz, C., Gelbrecht, J., Rennert, B., 2003. Treatment of rainbow trout farm effluents in constructed wetland with emergent plants and subsurface horizontal water flow. *Aquaculture* **217**, 207-221.

Schuyt, K.D., 2005. Economic consequences of wetland degradation for local populations in Africa. *Ecological Economics* **53**, 177–190.

Seck, P.A., Diagne, A., Mohanty, S., Wopereis, M.C.S., 2012. Crops that feed the world 7: Rice. *Food Security* **4**, 7–24

Sharma, K.K., Mohapatra, B.C., Das, P.C., Sarkar, B., Chand, S., 2013. Water budgets for freshwater aquaculture ponds with reference to effluent volume. *Agricultural Sciences* **4**, 353-359.

Sophocleous, M., 2002. Interactions between groundwater and surface water: the state of the science. *Hydrogeology Journal* **10**, 52–67. https://doi.org/10.1007/s10040-001-0170-8

Sørensen, J., 1978. Denitrification rates in a marine sediment as measured by the acetylene inhibition technique. *Applied and Environmental Microbiology* **36**, 139-143.

Stoorvogel, J.J., Smaling, E.M.A., 1990. *Assessment of soil nutrient depletion in Sub-Saharan Africa: 1983-2000: Nutrient balances per crop and per Land Use System. Volume 2: Report 28.* The Winand Staring Centre, Wageningen, The Netherlands.

Tittonell, P., Giller, E.K., 2013. When yield gaps are poverty traps: the paradigm of ecological intensification in African smallholder agriculture. *Field Crops Research* **143,** 76-90.

Twagiramungu, F., 2006. *Environmental profile of Rwanda.* European Commission and Government of Rwanda, Kigali. Available at: https://ees.kuleuven.be/klimos/ toolkit/documents/179_rwanda-env-profile.pdf.

UN-WWAP, 2009. *The World Water Development Report 3: Water in a Changing World,* United Nations World Water Assessment Programme. Paris: UNESCO and London: Earthscan. Available at: http://www.unesco.org/new/en/natural-ciences/environment/water/ wwap.

Unfinished Agenda, 1991. *Perspectives on overcoming hunger, poverty, and environmental management.* International Food Policy Research Institute, Washington DC, USA.

Usanzineza, D., Nhapi, I., Wali, U.G., Kashaigili, J.J., Banadda, N., 2011. Nutrients inflow and levels in lakes: a case study of Lake Muhazi, Rwanda. *International Journal of Ecology and Development* **19,** 53-62.

USEPA, 2000. *Principles for the Ecological Restoration of Aquatic Resources.* EPA841-F-00-003. Office of Water (4501F), United States Environmental Protection Agency, Washington, DC. 4 pp. Document number EPA841-F-00-003.

USEPA, 2005. *Protecting water quality from agricultural runoff.* USEPA Publication no. EPA 841-F-05-001. US Environmental Protection Agency, Washington DC.

Uwimana, A., van Dam, A.A., Gettel, G.M., Bigirimana, B., Irvine, K., 2017. Effects of river discharge and land use and land cover (LULC) on water quality dynamics in Migina Catchment, Rwanda. *Environmental Management* **60,** 496-512. doi: 10.1007/s00267-017-0891-7.

Uwimana, A., van Dam, A.A., Gettel, G.M., Irvine, K., 2018a. Effects of agricultural land use on sediment and nutrient retention in valley-bottom wetlands of Migina catchment, southern Rwanda. *Journal of Environmental Management* **219,** 103-114. https://doi.org/10.1016/ j.jenvman.2018.04.094.

Uwimana, A., van Dam, A.A., Irvine, K., 2018b. Effects of conversion of wetlands to rice and fish farming on water quality in valley bottoms of the Migina catchment, southern Rwanda. *Ecological Engineering* **125,** 76-86.

Van Asselen, S., Verburg, P.H., Vermaat, J.E., Janse J.H., 2013. Drivers of Wetland Conversion: a Global Meta-Analysis. *PLoS ONE* **8(11),** e81292. https:// doi.org/10.1371/journal. pone.008129.

Van den Berg, W.H., Bolt, H.R., 2010. *Catchment analysis in the Migina marshlands, southern Rwanda*. MSc-thesis in Hydrology and Geo-environmental Sciences, Vrije Universiteit Amsterdam and UNESCO-IHE Institute for Water Education, Delft, The Netherlands.

Van der Lee, G.E.M., Venterink, O.H., Asselman, N.E.M., 2004. Nutrient retention in floodplains of the Rhine distributaries in the Netherlands. *River Research and Applications* **20**, 315–325.

Van der Valk, A., 2002. *Methods for evaluating wetland condition: land-use characterization for nutrient and sediment risk assessment*. USEPA Publication no. EPA-822-R-02-025, U.S. Environmental Protection Agency, Washington DC.

Verdoodt, A., van Ranst, E., 2006. The soil information system of Rwanda: a useful tool to identify guidelines towards sustainable land management. *Afrika Focus* **19**, 69-92.

Verhoeven, J.T.A., Arheimer, B., Yin, C., Hefting, M.M., 2006. Regional and global concerns over wetlands and water quality. *Trends in Ecology and Evolution* **21**, 96-103.

Vromant, N., Chau, N.T.H., 2005. Overall effect of rice biomass and fish on the aquatic ecology of experimental rice plots. *Agriculture, Ecosystems & Environment* **111**, 153-165.

Wang, Q., Xu, J., Lin, H., Zou, P., Jiang, L., 2017. Effect of rice planting on the nutrient accumulation and transfer in soils under plastic greenhouse vegetable-rice rotation system in southeast China. *Journal of Soils and Sediments* **17**, 204-209.

Wang, S., Liang, X., Liu, G., Li, H., Liu, X., Fan, F., Xia, W., Wang, P., Ye, Y., Li, L., Liu, Z., 2013. Phosphorus loss potential and phosphatase activities in paddy soils. *Plant, Soil and Environment* **59**, 530-536.

Weissteiner, C.J., Bouraoui, F., Aloe, A., 2013. Reduction of nitrogen and phosphorus loads to European rivers by riparian buffer zones. *Knowledge and Management of Aquatic Ecosystems* **408**, 15.

Whitehead, P.G., Wilby, R.L., Battarbee, R.W., Kernan M., Wade A.J., 2009. A review of the potential impacts of climate change on surface water quality. *Hydrological Sciences Journal* **54**, 101-123.

Wollheim, W.M., Pellerin, B.A., Vörösmarty, C.J., Hopkinson, C.S., 2005. N retention in urbanizing headwater catchments. *Ecosystems* **8**, 871-884.

Wood, A., Dixon, A., McCartney, M. (eds.), 2013. Wetland management and sustainable livelihoods in Africa. Routledge/Taylor & Francis Group, London and New York.

Wood, A.P., 2013. *People centred wetland management*. pp. 23-64, In: Wetland Management and Sustainable Livelihoods in Africa (ed. by Wood, A., Dixon, A., McCartney, M.). Routledge/Taylor & Francis, 142 p.

Wood, A.P., Dixon, A.B., 2002. *Sustainable wetland management in Illubabor zone: Research Report Summaries*. The University of Huddersfield and Wetland Action, Huddersfield. Unpublished report.

World Bank, 2005. *Project appraisal document on a proposed global environment facility trust fund grant to the Republic of Rwanda for an integrated management of critical ecosystems project*. Report no. 30513-RW. World Bank, Washington DC.

World Bank, 2015. *World Bank open data*. http://data.worldbank.org/ (accessed 25.03.2015).

WWF, 2014. *Living Planet Report 2014: Species and Spaces, People and Places*. Worldwide Fund for Nature, Gland, Switzerland.

Yan, W.J., Yin, C., Tang, H., 1998. Nutrient retention by multipond systems: Mechanisms for the control of nonpoint source pollution. *Environmental Quality* **27,** 1009–1017.

Zedler, J.B., 2003. Wetlands at your service: reducing impacts of agriculture at the watershed scale. *Frontiers in Ecology and the Environment* **1,** 65–72.

Zucong, C., Guangxi, X., Xiaoyuan, Y., Hua, X., Haruo, T., Kazuyuki, Y., Katsuyuki, M., 1997. Methane and nitrous oxide emissions from rice paddy fields as affected by nitrogen fertilisers and water management. *Plant and Soil* **196,** 7–14.

Zuur, A.F., Ieno, E.N., Smith, G.M., 2007. *Analysing Ecological Data*. In: Statistics for Biology and Health (series ed. by Gail M, Krickeberg K, Samet JM, Tsiatis A, Wong W). Springer Science and Business Media, New York. 672 p.

Zuur, A.F., Ieno, E.N., Walker, N.J., Saveliev, A.A., Smith, G.M., 2009. *Mixed effects models and extensions in ecology with R*. In: Statistics for Biology and Health (series ed. by Gail M, Krickeberg K, Samet JM, Tsiatis A, Wong W). Springer Science and Business Media, New York. 574 p.

List of Acronyms

AIC	Akaike Information Criterion
ANCOVA	analysis of covariance
ANOVA	Analysis of Variance
APHA	American Public Health Association
ARA	Acetylene-Reducing Activity
DEM	Digital Elevation Model
DO	Dissolved Oxygen
EC	Electrical Conductivity
FAO	Food and Agriculture Organization of the United Nations
GPS	Global Positioning System
ICARDA	International Center for Agricultural Research in the Dry Areas
LULC	Land use and land cover
MEA	Millenium Ecosystem Assessment
MINAGRI	Ministry of Agriculture and Animal Resources
MINITERE	Ministry of Land
NICHE	Netherlands Initiative for Capacity Development in Higher Education
NISR	National Institute of Statistics of Rwanda
NPK	Nitrogen, Phosphorus and Potassium
NUFFIC	Dutch organisation for internationalisation in education
PSTA	Strategic Plans for the Transformation of Agriculture
SHER	Société pour l'Hydraulique, l'Environnement et la Réhabilitation
TN	Total Nitrogen
TP	Total Phosphorus
TSS	Total Suspended Solids
UN WWAP	United Nations World Water Assessment Programme
UR-CGIS	University of Rwanda Centre for Geographic Information Systems & Remote Sensing
USEPA	United States Environmental Protection Agency
WWF	Worldwide Fund for Nature

List of Tables

List of Figures

About the Author

Uwimana Abias was born on 08 September 1975 in Rwanda, Western Province, Nyabihu District, Bigogwe Sector, Kora Cell. From 1981 to 1990 he attended to the Primary Studies at Kora Adventist School. From 1990 to 1998, he did the secondary studies and obtained a high school diploma in teaching. In 1999-2000, he served as teacher at Kijote Primary School. From 2000 to 2005, he did his undergraduate studies and graduated in chemistry. In the period of 2005 and 2006 he served as teacher of chemistry at Groupe Scholaire de Muhura and Lycée de Kigali,  respectively. In the period of 2006 and 2007 he attended and graduated from the Master's Programme in Water Resources and Environmental Management at the National University of Rwanda (NUR). In 2008, he started working at NUR as Assistant Lecturer in charge of the Water Quality Laboratory. In 2009, he started his PhD studies at IHE Institute for Water Education, Delft, The Netherlands.

Publications

Uwimana, A., Nhapi, I., Wali, U.G., Hoko, Z., Kashaigili, J., 2010. Sludge characterization at Kadahokwa water treatment plant, Rwanda. *Water Science & Technology: Water Supply* **10.5**, 847-858. https://doi.org/10.2166/ws.2010.377.

Uwimana, A., van Dam, A.A., Gettel, G.M., Bigirimana, B., Irvine, K. (2017) Effects of river discharge and land use and land cover (LULC) on water quality dynamics in Migina Catchment, Rwanda. *Environmental Management* **60**, 496 - 512. https://doi.org/10.1007/s00267-017-0891-7.

Uwimana, A., van Dam, A.A., Gettel, G.M., Irvine, K. (2018) Effects of agricultural land use on sediment and nutrient retention in valley-bottom wetlands of Migina catchment, southern Rwanda. *Journal of Environmental Managmement* **219**, 103-114. https://doi.org/10.1016/j.envman.2018.04.094.

Uwimana, A., van Dam, A.A., Irvine, K. (2018) Effects of conversion of wetlands to rice and fish farming on water quality in valley bottoms of the Migina Catchment, southern Rwanda. *Ecological Engineering* **125**, 76-86. https://doi.org/10.1016/j.ecoleng.2018.10.019.

Presentations at conferences

Title	Type	Conference, Symposium or Workshop	Date	Place
Rehabilitation of nutrient and sediment retention functions in Migina catchment, Rwanda (PhD research methodology)	Poster	A2 SENSE (Environmental research in context)	January 2010	Apeldoorn, Netherlands
Assessment of land use effects on sediments and nutrient retention in Migina Catchment	Oral	International Scientific Research Conference under the theme " Energy Towards Sustainable Green and Affordable Energy	16-18 November 2011	Huye, Rwanda
Sediment and nutrient retention in valley-bottom wetlands of Migina catchment	Oral	Celebration of the World water day organized in partnership with the Ministry of Natural Resources and University of Rwanda	18 March 2013	Huye, Rwanda
Sediment and nutrient retention in valley-bottom wetlands of Migina catchment	Oral	Annual PhD Seminar	23-25 September 2013	Delft Netherlands
Effects of river discharge and land use and land cover on water quality and retention of suspended solids and nutrients in Migina catchment, Rwanda water quality	Oral	Meeting of Society of Wetland Scientists (SWS)	30 May-4 June 2016	Corpus Christi Texas, USA
Impact of conversion of wetlands on farming activities	Oral	Celebration of the World water day organized in partnership with the Ministry of Natural Resources and University of Rwanda	13-25 March 2017	Kigali, Rwanda

Acknowledgements of financial support

The research reported in this dissertation has been sponsored by the Government of the Netherlands through NUFFIC, the Netherlands

Netherlands Research School for the
Socio-Economic and Natural Sciences of the Environment

D I P L O M A

For specialised PhD training

The Netherlands Research School for the
Socio-Economic and Natural Sciences of the Environment
(SENSE) declares that

Uwimana Abias

born on 08 September 1975, Kora, Rwanda

has successfully fulfilled all requirements of the
Educational Programme of SENSE.

Delft, 28 November 2019

the Chairman of the SENSE board   the SENSE Director of Education

Prof. dr.   Dr. Ad van Dommelen

The SENSE Research School declares that Mr Uwimana Abias has successfully fulfilled all requirements of the Educational PhD Programme of SENSE with a work load of 48.8 EC, including the following activities:

SENSE PhD Courses

- Environmental research in context (2010)
- Research in context activity: 'Co-organizing Kigali Wetland Forum 8-12 July 2012 Wetlands: Wise Use, Smart Plans' (2013)

Other PhD and Advanced MSc Courses

- GIS and Remote Sensing Applications for the Water Sector UNESCO-IHE, Delft, The Netherlands (2011)
- Environmental Modelling, UNESCO-IHE, Delft, The Netherlands (2013)

Site specific training

- Summer School on Nanotechnology for Water and Wastewater , UNESCO-IHE, Delft, The Netherlands (2011)
- Laboratory water chemistry analyses, UNESCO-IHE, Delft, The Netherlands (2011)
- European SWAT Summer school, UNESCO-IHE, Delft, The Netherlands (2011)
- November GIS and SWAT workshop UNESCO-IHE, Delft, The Netherlands (2011)

Selection of Management and Didactic Skills Training

- Supervising two MSc students with thesis entitled 'Linkage between catchment characteristics, sediments and nutrients load' (2013) and 'Nutrients Loss and Retention in Rice Fields: case of Rwasave Fish Farming and Research Station, Southern RWANDA' (2014)
- Teaching a course component 'water treatment' of the module 'Water Treatment and Supply' in the MSc programme (2012-2014)
- Organization of the World Water Day, in partnership with the Ministry of Natural Resources and University of Rwanda, Huye, Rwanda (2013)

Selection of Oral Presentations

- Assessment of Land use effects on sediments and nutrient retention in Migina Catchment. Energy Towards Sustainable Green and Affordable Energy, 16-18 November 2011, Huye, Rwanda
- Effects of river discharge and land use and land cover on water quality and retention of suspended solids and nutrients in Migina catchment, Rwanda water quality. Meeting of Society of Wetland Scientists, 30 May- 4 June, Corpus Christi, Texas, USA

SENSE Coordinator PhD Education

Dr. Peter Vermeulen

T - #0113 - 071024 - C150 - 240/170/8 - PB - 9780367859732 - Gloss Lamination